GEESE

BREEDING, REARING AND GENERAL MANAGEMENT

BY

REGINALD APPLEYARD

("Backs to the Land" Broadcaster)

SECOND EDITION

PREFACE

I N writing this book I am not attempting to make it into a lengthy one full of long words and literary skill. What I wish to do is to place in the hands of the beginner, the general farmer, and the smallholder, a book which will be of help in giving a brief outline of the different varieties of geese and the purposes they serve; to give photographs of typical stock, and helpful suggestions in regard to Mating, Breeding, Incubation, Rearing and Management of young and adult stock.

There is no doubt many have taken up the breeding of geese as a side line, and for the purpose of eating-off grass. It is a great question what to do with grass these days, certainly the goose is a grass machine and will, at most times of the year, pick up much of its living in grass and herbage. Bearing this fact in mind and that a start in geese requires only a small capital outlay, I have not the slightest hesitation in advising those with suitable surroundings to try geese for a paying hobby or a profitable living.

I shall be pleased at any time to answer questions from any reader of this book.

REGINALD APPLEYARD

Ixworth,
Suffolk.

MADE AND PRINTED IN ENGLAND
AT
THE CHAPEL RIVER PRESS,
ANDOVER, HANTS.
2.45

Poultry Farming

Poultry farming is the raising of domesticated birds such as chickens, turkeys, ducks, and geese, for the purpose of farming meat or eggs for food. Poultry are farmed in great numbers with chickens being the most numerous. More than 50 billion chickens are raised annually as a source of food, for both their meat and their eggs. Chickens raised for eggs are usually called 'layers' while chickens raised for meat are often called 'broilers'. In total, the UK alone consumes over 29 million eggs per day

According to the Worldwatch Institute, 74% of the world's poultry meat, and 68% of eggs are produced in ways that are described as 'intensive'. One alternative to intensive poultry farming is free-range farming using much lower stocking densities. This type of farming allows chickens to roam freely for a period of the day, although they are usually confined in sheds at night to protect them from predators or kept indoors if the weather is particularly bad. In the UK, the Department for Environment, Food and Rural Affairs (Defra) states that a free-range chicken must have day-time access to open-air runs during at least half of its life. Thankfully, free-range farming of egg-laying hens is increasing its share of the market. Defra figures indicate that 45% of eggs produced in the UK throughout 2010 were free-range, 5% were produced in barn systems and 50% from cages. This compares with 41% being free-range in 2009.

Despite this increase, unfortunately most birds are still reared and bred in 'intensive' conditions. Commercial hens usually begin laying eggs at 16–20 weeks of age, although production gradually declines soon after from approximately 25 weeks of age. This means that in many countries, by approximately 72 weeks of age, flocks are considered economically unviable and are slaughtered after approximately 12 months of egg production. This is despite the fact that chickens will naturally live for 6 or more years. In some countries, hens are 'force molted' to re-invigorate egg-laying. This practice is performed on a large commercial scale by artificially provoking a complete flock of hens to molt simultaneously. This is usually achieved by withdrawal of feed for 7-14 days which has the effect of allowing the hen's reproductive tracts to regress and rejuvenate. After a molt, the hen's production rate usually peaks slightly below the previous peak rate and egg quality is improved. In the UK, the Department for Environment, Food and Rural Affairs states 'In no circumstances may birds be induced to moult by withholding feed and water.' Sadly, this is not the case in all countries however.

Other practices in chicken farming include 'beak trimming', this involves cutting the hen's beak when they are born, to reduce the damaging effects of aggression, feather pecking and cannibalism. Scientific studies have shown that such practices are likely to cause both acute and chronic pain though, as the beak is a complex, functional organ with an extensive nervous supply. Behavioural evidence of pain after beak trimming in layer hen chicks has been based on the observed reduction in pecking behaviour, reduced activity and social behaviour, and increased sleep duration. Modern egg laying breeds also frequently suffer from osteoporosis which results in the chicken's skeletal system being weakened. During egg production, large amounts of calcium are transferred from bones to create egg-shell. Although dietary calcium levels are adequate, absorption of dietary calcium is not always sufficient, given the intensity of production, to fully replenish bone calcium. This can lead to increases in bone breakages, particularly when the hens are being removed from cages at the end of laying.

The majority of hens in many countries are reared in battery cages, although the European Union Council Directive 1999/74/EC has banned the conventional battery cage in EU states from January 2012. These are small cages, usually made of metal in modern systems, housing 3 to 8 hens. The walls are made of either solid metal or mesh, and the floor is sloped wire mesh to allow the faeces to drop through and eggs to roll onto an egg-collecting conveyor belt. Water is usually provided by overhead nipple systems, and food in a trough along the front of the cage replenished at regular intervals by a mechanical chain. The cages are arranged in long rows as multiple tiers, often with cages back-to-back (hence the term 'battery cage'). Within a single shed, there may be several floors contain battery cages meaning that a single shed may contain many tens of thousands of hens. In response to tightened legislation, development of prototype commercial furnished cage systems began in the 1980s. Furnished cages, sometimes called 'enriched' or 'modified' cages, are cages for egg laying hens which have been designed to overcome some of the welfare concerns of battery cages whilst retaining their economic and husbandry advantages, and also provide some of the welfare advantages of non-cage systems.

Many design features of furnished cages have been incorporated because research in animal welfare science has shown them to be of benefit to the hens. In the UK, the Defra 'Code for the Welfare of Laying Hens' states furnished cages should provide at least 750 cm² of cage area per hen, 600 cm² of which should be usable; the height of the cage other than that above the usable area should be at least 20 cm at every point and no cage should have a total area that is less than 2000 cm². In addition, furnished cages should provide a nest, litter such that pecking and scratching are possible, appropriate perches allowing at least 15 cm per hen, a claw-shortening device, and a feed trough which may be used without restriction providing 12 cm per hen. The practice of chicken farming continues to be a much debated area, and it is hoped that in this increasingly globalised and environmentally aware age, the inhumane side of chicken farming will cease. There are many thousands of chicken farms (and individual keepers) that treat their chickens with the requisite care and attention, and thankfully, these numbers are increasing.

CONTENTS

ILLUSTRATIONS

CHAPTER I

MAKING A START

BEFORE making a commencement with geese, give the matter some thought and consideration, make enquiries and ask many questions—it costs nothing to ask them.

When starting, the points to remember are : What class of bird is most suitable for your purpose, and what scheme have you in mind taking into consideration the class of land, water and general situation? Obviously, if you live on a mountain side, with rough, poor herbage, you do not want exhibition Toulouse; the class of birds most suitable for such conditions would be really hardy Embden x Toulouse, Chinese or Chinese cross birds.

You may, perhaps, wish to go in for first-class pures such as Embden, Toulouse or Romans with the idea of breeding, exhibiting and selling stock birds, eggs and goslings at first-class prices. To do this it means commencing with the best, using skill and much care in mating, rearing, etc., and later on showing, advertising and generally telling and showing the public at home and abroad that you own, and have for sale, the real thing.

On the other hand, you may wish to go in for the production of goslings for sale at and from seven days to six or seven weeks of age—there is always a ready demand for such goslings, many people prefer to buy the required number of goslings to eat off and keep down spare grass. Thus there is generally a good sale for a young gosling at a profitable price to the producer.

There is also a demand and a good market for fertile goose eggs—at prices from one shilling to five shillings each if from very special stock.

If you wish to produce goslings to market when young try good class Embden x Toulouse, or grey and white

The above birds are the result of mating good class Embden and Toulouse, or Grey and White geese, to a White gander. Cross-bred birds are excellent for the Christmas market.

English geese, if possible heading the pen or pens with white ganders. Light coloured goslings for table purposes sell more readily than dark coloured ones.

If stock is required to produce large high-class Christmas Geese go in for good quality Embdens bred pure or an Embden gander mated with Toulouse geese. With the right class of stock you should get plenty of Christmas birds, weighing from 17 to 20 lbs. when dressed.

You may be in a district where the big goose or gosling is not popular, but where there is a ready sale for "Summer Goslings," "Michaelmas Geese" and small "Christmas Geese," of extra good quality; to produce such I would advise the Roman variety.

Make up your mind as to exactly what you require for your work before rushing in, spending good money and, later on, finding you are the owner of the wrong class of stock which you have to sell at a forced sale price and start again.

Having finally made up your mind, get into touch with some good breeder and state clearly your requirements. Ask him to quote you with full particulars for birds, carriage paid to your station, on two clear days approval against cash or deposit. Be prepared to pay a fair price for the right thing, and don't expect something for nothing.

Give the seller a clear idea of the purpose for which the birds are wanted, it is a great help to him. Suppose, for instance, you require stock to breed big Christmas Geese, it is highly probable that the breeder may have a big Embden gander that shows a little roughness in keel and gullet and is thus for sale at a reduced price. At the same time there may be some Toulouse geese of good size, but which fail in colour or in keel. These points are quite immaterial to you if your sole desire is to breed big table geese.

If you write a breeder simply saying you require a good sound pen of, say, Embdens, giving no details, you will naturally get a quotation for some of the best birds, whereas had you stated the fact that you wanted them to produce big table birds, you would give the breeder the chance of quoting a gander a bit rough in outline and some geese which show a few grey feathers—all at a smaller price yet most suitable for your purpose. Obviously a breeder of repute would not tell you he had Embdens with grey feathers if you asked for first-class Embdens!

When making a start you have choice of a number of ways. In their order they are (*a*) by the purchase of eggs; (*b*) by the purchase of goslings. (Both of these are all right but have their drawbacks in that when you have reared your goslings you have to sex them and buy unrelated ganders or geese to make up the pens, etc.); or (*c*) by the purchase of current season birds nine to ten months of age in October to December or early in the new year.

The last-mentioned is quite a useful way as you do know the age of the birds and that they have not been tested out as dud layers, layers of soft-shelled eggs or even non-layers. Again, you know the gander has not been tested as to fertility, and as ninety-eight per cent. ganders are fertile you do stand a chance of getting a fertile bird.

With such birds you would have good breeding stock for at least ten years with ordinary luck, and in the meantime be able to keep on as many young geese as wished from them, purchasing the male or males to mate with them.

Finally there is (*d*) the purchase of a good pen of over-yeared proved breeders, that is to say, birds that have proved themselves to be .good layers of hatchable eggs, fertile to the gander in the first season of laying. Such birds will cost more money, and it is only reasonable on the part of the buyer to expect to pay extra for a perfect and proved article.

As stated elsewhere first season geese are not always a success. If they lay eggs of good shape and texture, are fertile and hatch out even a small percentage of goslings, you will know that they should, in future years, prove reliable and be a success.

CHAPTER II

THE MATING AND SELECTION OF STOCK BIRDS

MANY may say there is not much need to write anything on this subject. Quite easy, just purchase a
gander and some geese and get right on with the job.
A large number do this, but only a few get results and suc-
ceed. Each week at certain seasons of the year I get letters
from would-be breeders who are troubled, so I think it
advisable to say a few words on the subject, for in any
scheme of breeding much depends on the mating. I
am not writing on matings for the purpose of breeding
special exhibition points, but on mating up a pen or flock of
geese in the hope of getting fertile eggs and sound rearable
goslings.

Haphazard placing together of a few geese with a gander
may possibly strike once out of six occasions, but even
where the object may be only the production of geese for
the market a sound system of selection must be followed.

When mating a new pen with current season birds,
seven to ten months of age, pick sound, healthy vigorous
birds typical of their breed. Handle each thoroughly,
looking and feeling for such faults as wry or roach
back, wry tail, rough in wing, and twisted or mal-
formed feet. Geese of good weight should be selected, of
course, making due allowance for how they have been fed.
Choose birds with plenty of length, width and depth of
body, refined in head, with bright, alert eyes—all these
features denote health and stamina. It is useless to keep on

a young bird lacking or wrong in some vital point; it will only cause annoyance and loss later on. The question of sexing birds will be dealt with in another chapter.

If you are using first-season geese, and somebody must use the one-year young geese, I would advise mating them with a second-season gander; failing such a gander, use a very early hatched bird. It is better, whenever possible, to use a second- or even third-season male with young females, as it at least gives maturity on one side of the mating. Thus the breeder has a much better chance of getting goslings sufficiently strong to get out of the shell, and, when out, rearable.

In choosing the gander, beware of the extra big, heavy-browed bird, inactive both in appearance and in movement. Certainly use a big bird if wished, he will increase size in the progeny, but do see that he is active, alert and looks capable of giving good fertility.

It will be found that first-season geese do not always prove a success from a breeding point of view, as their eggs do not always hatch out well, and even when they do the goslings are often inclined to be a little delicate and require extra care and attention for the first week or ten days.

Do not on any account get dissatisfied or part with the geese for this reason, provided, of course, that they have proved that they can lay good eggs. You may be parting with birds which in their second season, and up to ten years or even longer, may prove a continued success. A good breeder will never part with geese that have proved themselves, and, within reason, of course, there is no limit to the length of time that breeding geese can be kept.

Remember it is best to use a white or nearly white bird when heading any pen for the purpose of breeding table birds. He will give you a big percentage of white or light-coloured offspring which will pluck out a good colour for table purposes.

A " set " of one gander and three geese is the most useful mating, that is, in heavy varieties. With such varieties as Roman, Chinese, light breeds or crosses, four geese per gander proves a success. Five geese may, in fact, be used with Chinese as they are very active. It is possible to run two " sets " on the same range with success. It must

An evenly matched " set " of Embdens. In the opinion of the author a mating of one gander to three geese is the best for successful results.

be clearly understood that they must have a reasonably free range so that each gander is able to lead his wives about without always bumping into the other " set." In running two " sets " on the same range it will be found best, whenever possible, to use ganders which have been reared together and are reasonably friendly.

If you wish to run two, or even more, on the same range, give each gander his future wives and let them mate separately for at least a month, or even longer according to the time of year, as they will of course mate more quickly at certain seasons. When mating up pens, ring each bird of a pen with spiral rings—say, red for No. 1, white for No. 2, and so on.

When the " sets " are run together on the range there may be a few scraps and rows, but the birds will soon be

Here is a small batch of geese that have been specially produced for the Michaelmas demand. They are, of course, mainly crosses.

reasonably peaceful and in the future stick together in their own " sets." I have found mated pens will run together as a whole flock during late summer and winter and divide themselves up again early in the spring. Another remarkable thing I have often noticed is that, although they run as a whole in winter in the daytime, throughout the year each " set " retires to some special spot to spend the night!

This little revelation should effectively dispose of the oft-repeated assertion that if geese have any brains they seldom attempt to employ them! As a plain truth, I class them as being amongst the brainiest of all classes of domesticated fowls.

Generally there is no difficulty in introducing a gander to his geese so long as he meets all the geese at the first introduction. He will usually prove friendly and mate equally with the lot. The trouble is when you have a goose in a mated " set " which dies; the introduction of a new goose often proves an utter failure, as the old geese fight the new one and, worse still, the gander will not mate with her. In such a case the following method is advisable :

Take the gander right away from his geese; fix up a wire compound in some corner of the range or pen, as the case may be, and place the original geese and the new one into the wire compound; have the pen reasonably small, and it will be found that the geese will not hurt each other much when the male element is absent. Keep the geese restricted to the compound for three or four days or until they feed together and seem friendly. Now give them their liberty on the range, or in their pen, and once it is seen that they move about as a flock and are really friendly, return the gander, preferably in the evening. It will generally be found that the gander will divide his attentions equally amongst the geese.

And because the actual act of copulation is not observed, do not run away with the idea that no such act is taking place. In this one respect geese are exceedingly shy.

As far as possible matings should be made early in the season; they should certainly be completed by mid-January. By doing this the birds get used to each other, and quickly become settled and contented in their surroundings. All these points may sound and appear to be trivial, yet they are points which may be expected to count materially in the production of fertile eggs and, later, of sound, rearable goslings.

CHAPTER III

THE HATCHING AND INCUBATION OF EGGS

IN a later chapter on " Management of Breeding Geese," sufficient is written to prove reasonably to the reader that it is not a success to allow a goose to sit and hatch out her own eggs, at least, it is not from a commercial or pedigree breeder's point of view. Under special circumstances, and when dealing with ornamental varieties, which are kept under what might be termed a natural or next to wild state on a lake or large piece of water, it is of much interest to allow the goose to incubate, hatch and rear her own young. They are most fascinating, especially if their owner makes a habit of feeding them at a fixed hour each day, when they look forward to the visit, and, with care, quickly become hand tame.

Granted the goose is usually a wonderful mother, and rears a big percentage of the goslings hatched. For anyone interested in bird life there is no more charming or interesting sight to watch than a goose and gander with six or eight young geese that resemble golden balls or, in some cases, a beautiful shade of green or green and gold. The parents guard and care for them as if they were more precious than gold!

Without a shadow of doubt the way to achieve success, in my opinion, is to use broody hens. I like to have the nests made from turf (a grass turf cut in one piece, about 2½ to 3 inches in thickness) placed in the nest box and well beaten into shallow saucer shape—grass side upwards—and with a little cut straw on it. For nest boxes I much prefer

14

two-compartment orange boxes (three-compartment ones make the nests too small).

To prepare the box stand it top upwards and remove all but one piece of the top; now turn the box on its side, with the piece left on lowermost so that it can be used to hold

The hatch has commenced—young goslings show a brotherly (or should it be sisterly!) interest in the unhatched egg.

the turf in position; next take a piece of 2 × 1 wood, or about this size, nail it on to the top front edge—this is to hinge the doors on, and they are made from any odd light wood and framework, and sufficiently long so that, when hinged and in position, they will drop over the front bottom board and not into the nests. Then procure two ordinary bricks for each nest. These bricks are to lean against the door to keep it in position when the hen is sitting.

When you wish to liberate a hen, or the two hens, move the bricks, throw back the door on to the top of the box, using one brick to lay on the door and the other put in front of the nest to form a step for the hen. This is simple, most effective, costs very little and will last many years. Some I made for my own use very many years ago are still in service.

A range of sitting boxes as described in Chapter III. They are, as they should be, in a secluded, well-sheltered spot.

Any outbuilding, spare loose box, or poultry house is all right to put the boxes in, and they should be placed round the building to leave a space in the centre for the hens to feed, water and exercise. The floor should be kept sanded, or covered with sawdust or peat moss. Sand is best, as if a hen flutters the sand does not fly about.

Let us imagine we are going to put twenty eggs under hens. The eggs, which are already marked to their pen and dated when laid, are now re-marked very plainly at each end with the date when set under the hens. Four or five eggs go under a hen, sometimes six, but it depends on the size of the eggs and hen, also the season of the year. I do not like more than four normal sized eggs per hen in early spring when there is a chance of frosty, cold weather.

The hens are liberated each day at a regular hour, about four at a time. While they are off feeding the nests are examined and the eggs turned by hand. Long ago I came to the conclusion that a hen was never meant to turn a goose egg! One or two spare hens are always kept in reserve on dummy eggs, as we are again going against Nature in that a hen was not made to sit for from 28 to 30 days, the incubation period of a goose egg. After about fifteen minutes, any hen not returning is put back on to the nest.

Insect powder is needed from time to time; it should be sprinkled in and around the nests and on the hens.

The eggs are examined for fertility on the 9th or 10th evening by testing in front of a torch or even a candle, holding them in both hands between the first finger and thumb. If you cannot make a success of this, cut an egg-shaped hole in a stout piece of cardboard.

We now come to one of the advantages of sitting the eggs in groups and having each egg marked and dated. The clears will be taken away and the good, fertile eggs placed four or five per hen, thus often saving a valuable and much-wanted broody. Any hen which shows signs of going off or becomes restless should at once be replaced by a spare hen and another one procured in her place and put on the spare nest in reserve.

Having made many experiments with the hatching of goose eggs in incubators I give what is my honest opinion. I have done and can hatch out strong, sound goslings with the ordinary hot-air incubator, but results up to now are not a commercial proposition, and I do not advise the use

of incubators other than as follows, when they prove very
useful.

There is often much difficulty in procuring the required
number of broody hens early in the season and I have each
season tried the following method with succes: The eggs
are started in the incubator for ten or twelve days—temperature 102½-103° F. with water in the moisture trays from the
second or third day. The eggs are tested out and the fertile
ones go under hens. If I have hens on hand I move eggs
to them as soon as ever they can be tested out.   Each
season a 150-size incubator is especially kept for the purpose
of being used solely for young goslings.

I take any goslings from the hens as soon as they are
hatched, often, if in the evening, before they are quite out
of the shell, and place in the incubator. The machine has
the egg tray ready on which to put the eggs that are not
quite hatched and to put the goslings on which are not dry.
The nursery tray has a covering of clean wood wool which
is well pressed down in position. My idea in using wood
wool is to get over the difficult of any goslings getting
"sprawly legs" (see also chapter on Rearing).   Any
goslings on the egg tray, when dry, come to the front light
and fall on to the wood wool; failing this I put them down
by hand; if they are very lively and any trouble, I roll an
incubator felt and place it so that they cannot get from the
nursery tray to the egg tray.

Generally the machine is run at 104-105° F., as the
goslings are in the nursery tray and farther away from the
heat. On some occasions, when the goslings have to be
kept in the nursery for a good time, I lower the lamp and
have a special length of wood with which I prop open the
incubator door an inch or so.   Later on the goslings go
back to the hens.

If the weather is hot and dry I sprinkle—often pour—
water down the back of the broody boxes. This goes
under the boxes and is absorbed by the turf (another good

reason for using turf), also sometimes I sprinkle the eggs when near " chipping." If you do sprinkle the eggs, do it just before the hen is returned to the nest, and use warm water. In very hot, dry weather, with later eggs, you can even wet the breasts of the hens as well as sprinkling the eggs.

As with ordinary fowls cross-bred geese are far and away the best sitters and mothers. The author prefers ordinary hens to geese as broodies.

In the hatching of goose eggs you will often find that a gosling will get one hole in the shell and with its beak out— I mean well out, not just the tip in view. A gosling in such a position hardly ever gets out of the egg, unless it receives some attention, as it cannot get the beak back, and if it cannot get the beak back it cannot get any leverage to continue the good work. Push the beak back—take a piece of goose egg-shell as nearly the proper shape as possible, and stick it over the hole with gum; failing gum, use white of egg, and failing all else buy and use surgical plaster—it works; I have tried it even with pigeon eggs, using stamp paper.

I much prefer to do this rather than attempt to " hatch " the youngster myself by breaking away the egg. In this particular, geese are very similar to ordinary fowl—a humanly assisted " hatch " is seldom successful in the long run.

Incubating and hatching geese is one of the most interesting and engrossing tasks of the year, and, as long as I have been in the business, I continue to look forward with almost childlike impatience to the commencement of a fresh hatching season. Rearing and managing the youngsters is, of course, the subject of another chapter.

CHAPTER IV

DISTINGUISHING THE SEXES

ISTINGUISHING the sex of geese has, in the past, caused much worry, annoyance, and, in many cases, loss of time and money. The question crops up hundreds of times each season; all sorts of answers are given, many of which are helpful, such as : " General masculine appearance and outline, size of head, length and thickness of neck, the stretching out of neck and hissing in the male bird.... That the flock of geese be herded in a corner, a small log procured, etc., when the ganders will come forward on the outside of the flock .... A double pouch, or bags between the legs in the goose and single pouch in the gander."

Another test is to take the bird you wish to distinguish, place it out of sight, but in hearing of the other birds; they will then call and answer each other; if you have a musical ear it is quite easy to distinguish the sex! Another method, and one which is most certainly positive, is to keep the birds over the year; those that lay are geese! Most expensive advice, especially if you get a big percentage of ganders! It rather reminds one of the story about the farmer who had an arable field, very subject to rampant growth of colt's-foot. After trying many things, it had the farmer beaten; the village lunatic had given the matter much thought. He stopped the farmer one morning and guaranteed, for the sum of two-and-sixpence cash, to give a permanent cure. The farmer paid, receiving the following advice : " Flag the field all over and cement up the nicks." Certainly a cure, but again very expensive.

Year in, year out, people give all these methods a tria
with varying results. I am giving what is, in my opinion
and what I have found to be, the most reliable and provec
method of distinguishing the sex of geese. First, let m
say the birds must be matured—say seven months or older
the older the better. Secondly, gentleness and care must b
used. By this method, in the right hands, it is possible tc
state the sex of 90 per cent. in young matured birds and 10c
per cent. in old birds. Some little practice is necessary anc
no force must be used.

Having caught the bird you wish to discern, lay i
gently but firmly on its back (it is helpful if you hav
an assistant), and kneel with your left knee on what are th
nearside flight feathers (on no account kneel on the wing
bones on which the long flight feathers grow). Pass th
bird's head and neck through your fork and between th
bend of the right leg (also in kneeling position). You
assistant will hold down generally and care for the othe
wing and leg.

Turn the feathers and fluff away from the bird's genita
organs. You should now have exposed to view a fleshy par
about the size of a shilling (shown in diagram A and B)
With the first two fingers of one hand, use a gentle but firn
outwards and downwards pressure, then, under ordinar
circumstances, and with matured birds, you should soor
know the sex. By referring to diagram my explanatior
will, I hope, be made clear to the reader. In the diagran
I have tried to show the organs of each sex when viewec
from above in A, male, and B, female.

In each case, of course, when the pressure has exposed o
partly exposed the parts—I say partly exposed, as in A th
operation has not exposed the male organ—the operation i
really successful if it is possible to expose the male orgai
(A2). But it is not always so, and we must come to othe
helpful aids, such as the fleshy part in the male, which i

much firmer than in the female, and if felt or even viewed from the other side is inclined to be convex (A1).

Without going into further details it is hoped the reader will grasp my meaning with the aid of the diagram. In the case of the goose (female) it will be found that the fleshy part is much softer, a darker shade, nearly red (generally) when viewed from the side inclined to be concave (B). The small sketches should make clear what is most difficult to explain in writing.

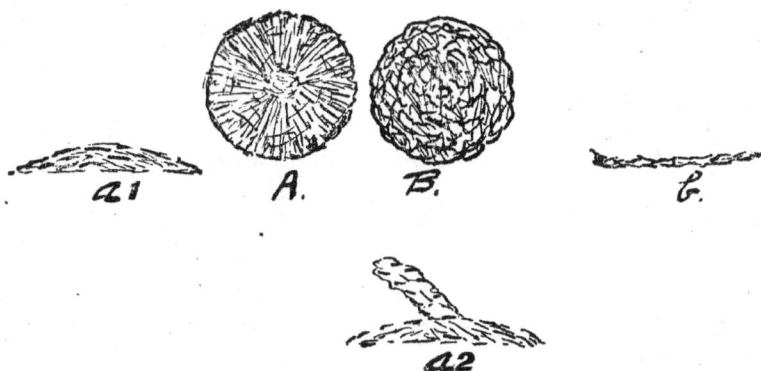

Sexing Geese : Explanatory diagram.

In these examinations, on no account must any force be used. When handling the birds let every movement be careful and gentle; this helps to success. I do not say that this method is infallible, but I do say that with experience and practice it can be made to give 90 per cent. correct sex. When breeding pure varieties of geese, such as Embdens, Toulouse or Chinese, it is fairly easy to distinguish the sexes in that there are many visible signs in outline (often size), etc. Even then, when breeding pure varieties, we get exceptions, such as a very big female or a small male, each with every appearance of their opposite sex. In such instances the method given proves most useful.

CHAPTER V

THE MANAGEMENT OF BREEDING STOCK

For the purpose of going into the question and details relating to the management and care of breeding stock, I think it will make it clearer to the reader to imagine that we are dealing with two pens of birds, say Embdens and Toulouse, each an over-yeared pen of gander and three geese—each lot, of course, on its own ground or pen.

With these two imaginary pens in our mind's eye I will now as briefly as possible touch on all the main points of management likely to help in procuring what is required, *i.e.,* eggs and goslings.

During the back end of the year these birds will be fed on good foods in sufficient quantities to keep them in good, hard, healthy condition—on no account let them get fat, and on the other hand, do not let them get into a lean and weak condition. Follow Nature as much as possible. In a wild state the goose does not, during the dead months of the year, live on the fat of the land. As a matter of fact it is just the opposite; it has to work hard to procure a living plus keeping a wary eye on the pot hunter and punt gunner. As the new year arrives and days go on, the wild goose finds more and more animal foods, young green grasses, weeds, etc., which gradually bring it into breeding condition.

I have always found that the most successful of small livestock keepers are those who make a close study of Nature. While the domestication of fowls, and animals, compels us to adopt some practices that are entirely against Nature, there remains no reason why we should do everything as a habit and deliberately flout the natural order.

Early in the new year the breeding stock will be put on a breeding diet and fed at regular times—grain feed of wheat and maize each morning (half of each and, as the weather becomes warmer, less maize).   Use your own judgment, giving just as much food as the birds will eat and leaving them looking for more.   Each evening give a good feed of wet mash, mixed from ordinary laying mash and fed in a crumbly state. Give sufficient so that they clean it up.   It is essential that this mash contains ten per cent. of white fish meal, or its equal in meat and bone, or whale meat, preferably guaranteed 60 per cent. albuminoids.

The owner must use his or her judgment as to the quantities fed.   A lot depends on what natural food the birds can pick up on their range or pen.   The great thing is to get the birds into perfect breeding condition and bloom by mid-February—not too fat, but in a vigorous active state.

The laying shelter will be fixed up or, if already on the place, reconditioned and surroundings generally cleared up and nesting material such as straw or rushes supplied. Oyster shell or limestone grit must be supplied *ad lib.*; this latter I prefer to place in small heaps on the grass near the shelter or in close proximity to where the birds feed.   For some reason the geese seem very shy of it when placed in a container, and, seeing that the taking of some form of shell-making material is absolutely essential, one must use every effort in getting them to eat it.   If necessary mix some fine shell or limestone in the mash from time to time.   A lot will depend on the district and what shell-forming material they can pick up.   The owner or person in charge must do what is considered best and work on experience.   One thing is certain, soft-shelled goose eggs are a most useless and annoying thing, and are to be avoided.   I have found limestone grits and limestone flour most useful.

Everything possible must be done for the breeding stock to ensure that they are comfortable and happy in their surroundings.   Under normal circumstances, but, of course, depending on the weather and management and the age of the birds, the geese should come into lay in mid-February or early March.

The great thing to guard against in breeding geese is to see that they are not too fat before commencing to lay. If too fat they will give you lots of double-yolked or soft-shelled eggs. Once get a goose laying these big eggs and it becomes a habit, regular and most annoying. When geese are in full lay I much question if one can overfeed them if feeding the right class of foods.

Certainly good and proper feeding when in lay will help to prolong the laying period. What is wanted is lots of good-sized, well-shelled eggs at the right season of the year. Without these you cannot get goslings, and without goslings your breeding pens are a failure. If you have not a reliable party to feed and attend to them feed the birds yourself.

Having got your geese in lay, the next question is how to store eggs until required for sitting purposes. The following is the method I have practised for many years, and it gives very satisfactory results. The eggs are collected each day; in frosty weather they are gathered as quickly as possible, although the goose generally covers up the eggs after laying, and they seem to take little harm in cold weather if not too frosty.

Handled with care, they are at once marked to their pen and dated (whenever possible and when pedigree breeding they are marked to the goose that laid the egg). If dirty they are placed in warm water and carefully cleaned, if necessary using a scrubbing brush. By making them thoroughly clean it helps very much later on when testing for fertility. Having washed and dried the egg or eggs, see that pen and date marks are distinct. Store in a box on clean sweet chaff or bran and place on their sides, which I think most will agree is the natural position in which to store any egg. Turn each day until required to " sit."

Although I have stored eggs of geese for what might

appear to be terribly long periods, I agree with the often
expressed opinion that the fresher the eggs are when put
down for incubation the better will the hatch be. Storing
eggs indoors is a vastly different proposition from the goose
storing them herself in, say, an undetected nest.   Her

Here are a couple of broody geese in a coop such as is briefly described below.

regular visits to the nest keep the older eggs as " alive "
and as fresh as those that are only a day or two old.

The box should be kept in an even temperature and in a
dark place; failing a dark place cover the box with a clean
piece of sacking.

We will now imagine that the geese have laid a number
of eggs and that one or more birds show signs of broodiness.   This will be quickly noticed by an observant

attendant as the broody goose will remain on the nest for hours after laying. Immediately it is noticed that the goose clings to the nest without laying an egg, act, and act quickly—the quicker the better for our purpose. The idea is to quickly break the goose from broodiness and get her into lay again as soon as possible.

Procure a large, roomy, bottomless coop or box, of course with bars in front, place it in the pen on a clean, level grassy place and in this place the broody-inclined goose. Ample food and water should be supplied and the coop moved on to fresh grass each day.

One of the best and cheapest class of coops for the purpose is a large packing case, made from laths or riven wood. These empty cases can be bought very cheaply at most stores. Just take off the top bars, turn upside down with the top on the floor, slip in the goose or geese, and see they have room to move about and to eat. Have in mind that, although you have placed the goose in prison, she is not there for hard labour and punishment; on the contrary, our purpose is to prevent her getting back on to the nest, yet to get her back into condition to lay again. Thus plenty of food, water and a move each day on to fresh grass are necessary.

On no account take the goose out of sight and hearing of her mates; do so and you may find that the gander will pay her no attention when returned to the pen, or, worse still, the other geese may knock her about. Cooped in the pen and liberated when you consider she is over the broodiness, she will soon come into lay again and give you another batch of eggs. Collect these second batches of eggs, mark, store, etc., as before. The great idea to my mind is to use the goose as a producer of eggs, not as an incubator. As an incubator she generally makes a mess of things, giving very poor results. I will explain what I mean; the reader must then use his or her own judgment as to whether to use the goose to sit and hatch eggs.

Use your goose as a broody for its own eggs, and you get no early goslings! Why? Well, it will take the average goose, say, quite three weeks to lay the number of eggs she can cover (say 12-15, depending on size of eggs and size of goose). Even then, when she has laid the required number, she may not be broody, but lay on up to 20-30 eggs, by which time your eggs are stale.

Even then give her eggs to sit upon and generally speaking you get no second laying of eggs. ·If she does lay, which is seldom, the eggs are generally unfertile, or they would give you very late goslings. Granted there is no more delightful sight than a goose with eight or ten golden goslings on green grass, but generally speaking you do not get the eight or ten goslings, as the goose hatches out a few of the fresher eggs and walks away with two or three goslings, leaving any other goslings in the eggs to perish. Geese are grand mothers and will generally rear any goslings they hatch out, but from a business and commercial point of view it is foolish to use the goose to hatch out her own young.

It is a very good plan to keep a chart and mark up all eggs laid by the different pens or pen. It gives one a good idea at the end of the season how many eggs were laid and is most interesting and helpful in building up a strain of good layers. Whenever possible (by shape of eggs, or separate nest), one should try to keep individual records.

CHAPTER VI

REARING YOUNG STOCK

To the average person the care and handling of young live stock gives much joy and pleasure in life. In the case of goslings we are dealing with what I always consider one of the most interesting, fascinating and intelligent of young stock. They come into this great big world so beautifully formed and finished off in every feature; real little men (and ladies!), and very independent, without any signs of fear for human beings. Each spring I look forward to the arrival of my first lots of goslings. With pedigree stock of such varieties as Embden and Toulouse the earliest goslings I remember were hatched on March 5, exceptionally early. The most general date to have goslings arriving is the end of March or early April.

In April all Nature is just getting well awake—short crisp green grass, and without the latter on which to coop your goslings they are never really content and happy. From a day to six weeks of age a well-bred gosling is most interesting to the owner or attendant who makes a hobby of his work, and you cannot make a great success of any class of work unless you are keenly interested in it.

Now for a few words on rearing goslings, with hens and with artificial mothering, *i.e.,* foster mothers or hovers. First, it is absolutely essential to success that on no account should any goslings be taken out of the nest or incubator until they are thoroughly dry, well fluffed out, and absolutely sure on their legs; see that they can stand up and run about before putting out under the hens or to the foster mother. Do not be in a hurry; they can well do without any food for thirty-six or even forty-eight hours if necessary after hatching.

Let us for the moment imagine that the first twenty eggs put down gave us fifteen sound goslings hatched by hens; they are to be reared in coops with the hens.

A handy type of broody coop and run for the use of hen and young goslings (see Chapter III). This should not be confused with the coop described in the previous chapter.

Procure three roomy, sound coops—each with a good front and loose floor. With the coop we also require a run, either a wooden one with wire top or about two or three yards of small wire netting with some small stakes with which to form a run in front of the coop. A water container (which must on no account be a wide shallow one such as a saucer, which is useless as the goslings only paddle or sit in it, lose all the water and at the same time wet

B

themselves, become bedraggled, wet the coop floor, get a chill, which results in pneumonia and death) must be provided.   I have tried most things and have found that any old tin about four inches in diameter cut down to a height of about three inches (a small-sized Lyle's Golden Syrup tin cut to required height) proves ideal.   Best of all is a small glass tongue jar.   If the water vessel is the correct height the goslings can easily get a drink and immerse their heads, yet cannot get in and wet their bodies.   If the container is too big I drop a good sized stone in or two or three smaller ones.

Provide for each coop a clean rough sack.   Take the coops on to the rearing ground, placing them on clean level ground where the grass is short and sweet.   If the grass is dry, place floor and sack on the coop top, the sack under the floor so that if it rains the sack will be kept dry.

Give each hen five goslings.   If they are of different colours, some dark and some light, see that each hen receives some of each.   If you do not do this, you will find later that some poor innocent grey gosling will walk into a coop which is occupied by a hen with light-coloured youngsters and in most cases will get killed or badly mauled about.

Place the hen in the coop, then put the five goslings under her—rear the coop front slightly to darken it a little. Leave for a time, then return and carefully note if the hen has taken them, and is properly brooding.   I like, when first placing goslings under the hen, to let one move about in front of her just to see if she is friendly with it.   The hens should have been fed and watered in the brooder house before cooping.   Very young goslings easily get on their backs if any holes are present or uneven ground is used. Keep your eye open for this—once on their backs, unless quickly righted, they soon die.

Now feed and water—placing the water pot just inside the coop.   Sprinkle a little broken green grass (break small by tearing with fingers) on the water, make some of the food into a small ball and place it in front of the hen, she will break it up and in falling to pieces it will help to attract the goslings. Do not get worried if they won't eat or if they eat very little, they will do so when ready.

At night place the floor board (coop bottom) on the ground, on this place the sack, any over the ends or back of floor fold under, see that the sack is level and well stretched. On to it place the coop and put in the hen and goslings. Once the goslings are under the hen and happy, put the coop front in position for the night.

The reasons for using the sack are as follows: If you leave the hen on the grass, especially early in the season, a certain amount of moisture will rise and make the whole wet and damp. In a wild state the mother would brood her young on some dry spot and under shelter, at the bottom of a tree, or under a shrub with reasonably dry litter beneath it. Another reason for the floor is for protection against rats and other vermin.

. To my mind the sack is absolutely essential to success in that a coop floor is absolutely against Nature, and a young gosling cannot get a grip or foothold on a smooth slippery wood floor, thus many otherwise good goslings become "sprawly legged." By this I mean the legs and even the thighs go out sideways and the poor youngster tries to get about on its stomach. It is next to impossible to put them right once they are allowed to get sprawly. The sack does the trick, the babies can get a grip on it and, what is more, it makes the coop bottom nice and dry and comfortable for the night. Use a clean sack each night, or the old one hung out to dry and air ready for use.

Another successful way, when the ground is dry, is to place a piece of one inch mesh wire-netting on the floor with the coop on it. This is most efficient if you pick a level spot and see that the wire is flat on the ground. The coops are moved each day, at least once, but often two or three times. I say once a day for certain, as this move does away

B*

with the question of the goslings getting any stale left-over foods.

We very quickly arrive at the stage when the goslings make such growth that they find difficulty in getting through the coop bars. We can generally get over this by running three or four lots of goslings as a group by drawing a bar from each coop and allowing the hens and goslings the run of a large wire compound, say 25 yards by 18 inches high, with 2-inch mesh netting staked in position round the coops.

Later on, as the goslings grow bigger and of such size that it would be unwise to fasten them in a small coop at night owing to lack of air, it is best to run the group at nights into a 6 × 4 × 4 house with a sound floor covered with litter and with adequate ventilation and fresh air. When they get bigger, some feathers on their backs, and are reasonably safe from vermin, the house door may be firmly propped open, to allow them to please themselves about going to bed. Later they will sleep in the open and, once feathered, they require no house and are not much trouble.

I do not propose to say much on the subject of artificial rearing for the simple reason that up to now no great success has been made in artificial incubation. Generally speaking it is quite easy and simple to rear goslings under hens, and in my opinion hens prove the best and most reliable in the hatching of goose eggs. There is no need to provide brooders or foster mothers when you already have a brooder in the broody hen, who is well deserving of a rest and good food after sitting for thirty days.

As already indicated in a previous chapter, the rearing season is probably the most engrossing and interesting part of the year, and I can think of no more pleasing, and sometimes exceedingly amusing, sight than a broody hen with her full complement of young goslings.

Goslings can be most successfully reared in the ordinary

type of brooder or foster mother, such as the Gloucester brooder—brooder compartment and covered run all under one roof. There is just the difficulty of getting the youngsters out on the grass. This I have got over for the first few days by cutting thin turfs of green grass to place on the brooder floor until they learn to run in and out on their own. Run at a temperature of 90 degrees for the first four days then at 80 degrees for a week and gradually lower until outside temperature is reached. With just a little extra attention for the first few days, until they learn to return for warmth after feeding, they will be no more trouble than chickens and certainly not so much trouble and worry as turkey chickens reared under similar conditions.

For the very small quantity of hand-fed foods eaten by a gosling during the first five or six weeks, I have always considered it a good policy to feed well on the very best foods it is possible to obtain. At first little and often, and early and late, gradually reducing the feeds to three per day.

Fed well and on first-class foods for six weeks you should have built a good frame and have a sound strong foundation for the future. As much as possible keep on short sweet grass; long coarse grass is no good for baby goslings. Grass which is short and young is the thing; such grass is a wonderful food.

For the first few days feed fine biscuit meal, scalded and dried off with Sussex ground oats. After four days dry off the biscuit meal with any good first-class chicken mash, preferably containing two per cent. cod liver oil and minerals. I generally sprinkle a little fine limestone grit or flour on to the mash and mix it in. Any double-yolked, cracked or thin-shelled eggs are broken and go into the feed raw, placed on the scalded biscuit meal and well stirred in with a spoon before the whole is dried off with meal. Later, when the goslings are about a week old, I mix a little No. 1 chick grain into the mash.

Another thing I like very much is a clean sound sample of wheat put through an oat crusher. In other words, crushed wheat. Half fill a bucket, scald and allow to stand for two or three hours with a sack over it, then use some in place of the biscuit meal in the mash. It is the whole wheat and nothing but the wheat, reasonably cheap, which helps to bulk the mash and is much relished by the goslings. Later on add some dry whole wheat and cracked maize in the mash. Once feathered they must have at least one grain feed per day, on which they will do well without other foods if they have plenty of grass to eat.

Never overfeed, if anything feed on the sparing side, especially when grass is abundant.

In rearing there are one or two essential points to watch : Provide shade from sun in very hot times; give plenty of water; never shut up a group of wet goslings in a poorly ventilated damp or dirty coop or house; rear really well for first five to six weeks. It never pays to stunt a gosling for life by trying to save a few pence in the early stages.

This chapter has, I know, been revised at a time when war-time rationing is worrying every class of small live-stock keeper, and it will most certainly be read long before matters return to normal again. But even under these trying conditions we must offer our young growing geese the best of those foods that are available to us.

CHAPTER VII

HOUSING OF GEESE AND GOSLINGS

PERSONALLY I have no use for houses for geese. In my opinion we have in the goose a really hardy unspoilt bird, so why pamper it and thus make it less hardy? Having a bird which does well unhoused, why go to the unnecessary expense of providing housing? There may be exceptions, of course, and the owner must use his own judgment. Much depends on the class of bird kept and the district and situation in which they live.

I am positively certain that were I to offer houses to my geese I should still find them sleeping out under the conditions that are more natural to them, and which they obviously prefer. There are, however, exceptions to this rule, as the reader will find later in this chapter.

Obviously an absolutely exposed position, without any hedges or shelter from all the winds that blow, may cause one to erect some form of housing or shelter. In places well sheltered by natural hedges, shrubs and trees, housing, to my idea, is unnecessary. Many may be fortunate in having a spare loose box, an old shed, or open-fronted building, if so, certainly make use of it—it should be well littered with straw, bracken, old grass or rushes, and it will offer the birds shelter if they wish to make use of it. In districts troubled by foxes they may be closed in at night, but it is a good and very hungry fox which will tackle a gander or goose! Such sheds are very useful during the laying season.

In the early spring I like to erect shelters under which the geese may have their nest. Any old shelter does well. A few wattle hurdles staked in position with a thatched roof, sheets of galvanized iron on their sides against a boundary fence and staked into position, or the same with a wattle hurdle or an old door, do really well for the goose.

Make a proper shed, if you wish, of framework and new faggots for the sides with thatched or galvanized iron roof. Make use of material you have handy, or material that is cheap and easily procurable. For very little outlay you can make quite nice-looking rustic shelters. After all your trouble you may be surprised to find the geese will make their nests in the open—in a clump of grass—or under a bush, etc. As I say, they have not much use for man-made houses or shelters.

Once the goslings are feathered there is not much need for houses. Goslings fully feathered, taking the very early ones at say six to eight weeks of age, should be finished with coops and 6 × 4 × 4 ft. houses. Early goslings hatched at the end of March would be nicely feathered by the end of April or early May and thus, having quite good weather, require no house in which to sleep.

They are healthier, happier and better unhoused and at liberty, ready in the early mornings to get their breakfast of green grass just like they would in a wild state. If housed, plenty of room and air are necessary with very early liberation each morning. Thus, I say, house only when absolutely necessary.

Many must use a house or houses for their adult breeding stock. In such cases a useful size of house for a gander and three or four geese would be about eight feet long, five feet wide, six feet high in front and four feet at the back. A good wide door should be provided, and most of the front should be wire netting. For a flock of, say, two ganders and eight or nine geese a house about 12 ft. by 6 ft. or 7 ft. would prove useful.

Very useful, quickly erected houses can be made by the

use of 2 × 3 and 2 × 2 inch wood and corrugated-iron sheets. The floor can be of cement if wished or of rammed chalk; put in wet and well rammed and beaten level it will set like iron. Such floors should be kept well bedded with straw, rushes, bracken or some such material.

Whichever type of house is employed, pay particular attention to the entrance. Popholes are, of course, dispensed with, and entrance is through the door. Never expect a goose to jump over a threshold; it will most probably

On many farms shelters such as this are erected for the use of geese on range. While not absolutely essential they are often appreciated during hot spells of weather.

stumble over any object in its way. For this reason keep any thresholds low to the ground or provide a suitable run-up.

It is a good plan to paint or give a coat of black bitumen varnish to the under side of the iron sheets before nailing on to the woodwork. This will, in some ways, do away with the drip from the roof and will certainly give a much longer life to the sheets. If you do erect a house, choose a high, dry, sheltered, quiet position for it.

CHAPTER VIII

MARKING AND RINGING

TIMES out of number one is asked "Can you tell me how to mark my goslings or adult birds so that I may know them, etc.?" There are many ways of marking goslings and adult geese so that there is no difficulty in distinguishing them if, for some unknown reason, they get mixed or are run as a flock after the breeding season.

We will commence at the beginning and deal with the question of marking goslings. Just for the moment imagine we are dealing with the progeny of three different pens of Embdens, eggs sat under hens; all eggs marked to each pen, say Embden No. 1, No. 2 and No. 3.

When the eggs are due to hatch out they are, as much as possible, separated in lots from the different pens, and given to the different hens, thus, say, two hens will be hatching pen 1 eggs, another hen pen 2 eggs, and so on.

The best method I know to mark goslings is as shown in sketch herewith. Placing the goslings on a clean, smooth piece of wood, cut a V shape piece of the web out with a sharp knife, or a safety razor blade or a small pair of scissors. Personally, I don't like scissors as they bruise the web as well as cutting it. Do not be afraid of the operation, it does not appear to hurt the baby, nor does it in any way prove harmful.

Now see diagram of a pair of gosling's feet. You are looking at a near-side leg and foot and an offside leg and foot.

The near-side leg of any bird is the left leg when you

PENMARK	PEN NO	GOSLINGS 1933 NUMBERS MARKED...
O S N S	NO.1. EMBDEN	////////X
	NO.2. EMBDEN	/////
	NO.1. TOULOUSE	//////X//
	NO.2. TOULOUSE	////
	NO.3. TOULOUSE	//////X
	NO.1. ROMAN	//////////////
	NO.2. ROMAN	//////////XII
	FLOCK ROMAN	////////////////
	NO.1. PAIR CANADA	////
	NO.1. CHINESE	////////////////
	NO.2. CHINESE	///////////////
	NO.1. EMB: & TOUL:	////////XII
	NO.1. BUFF	///////XI

X DENOTES DEAD

A record card showing the output of each pen of geese should, of course, be kept. The above is an excellent example.

are standing at the left side of it and looking in the same
direction as the bird.   The offside leg is the right-side leg
when in same position.

I especially wish to make this clear as later I shall write
of such as near-side inside web, near-side outside web and
offside inside web, offside outside web.

In the incubator room or broody house or in a pocket
book you will keep a record of goslings hatched and marked

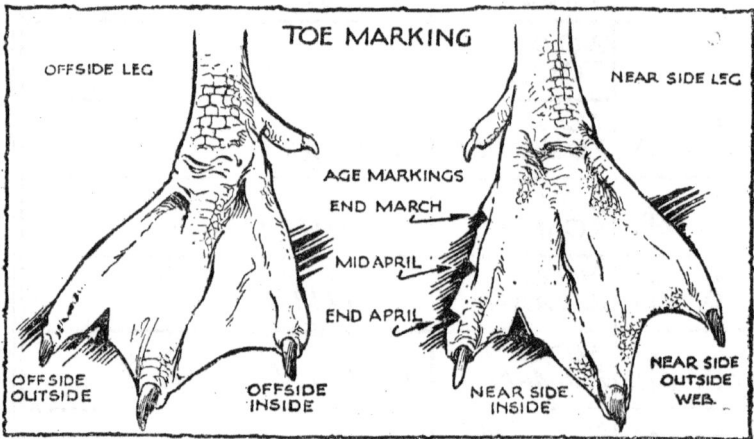

TOE MARKING

OFFSIDE LEG

NEAR SIDE LEG

AGE MARKINGS
END MARCH

MID APRIL

END APRIL

OFFSIDE
OUTSIDE

OFFSIDE
INSIDE

NEAR SIDE
INSIDE

NEAR SIDE
OUTSIDE
WEB

as shown (i.e., record card).   If you mark all goslings
hatched, and each time you mark one place a stroke opposite
the pen on the chart, you will later find you will know at
the end of the season exactly how many goslings you have
had from each pen of birds.

Any deaths of marked goslings can be marked off by
putting a stroke across the first stroke as shown on record
card thus : x.   The great idea is to finish the season with as
few x marks as possible.   Naturally, some of my readers
only keep birds purely for table purposes, or only keep one
pen—thus not of much value to mark.   I mention this
marking for use of those interested in pedigree stock and

who keep a number of pens and produce their own future stock birds.

Having marked our newly hatched goslings and got them on mother earth they will grow quickly, so we come to the question of ringing.

To the breeder of pedigree stock I would certainly advise the purchase and use of Conference rings; they are made from aluminium, each dated with year and a number. Two kinds are issued, Waterfowl Club rings, and Poultry Club rings. The first may only be sold to members of the Waterfowl Club, the latter to anybody.

It is difficult to give the exact age at which these rings should be slipped on the leg of a gosling, as much depends on the breed and strain, how well-grown for age, etc.

It is up to the owner to get them on at the right time. If you put them on too early you will lose some, leave it a few days too long and you cannot get them on at all. Just one thing when ringing a bird with a numbered ring— place it on with the number upside down when the bird is in a standing position. At any time later in life when handling the bird it will be found that when you are holding the bird under your arm the number is the right way up for you to see. If placed the other way you will have to almost turn the bird upside down so that you may read it. Later on it is a good plan to ring all the birds with spiral rings (big size Nos. 9 and 10).

A good idea is to use red for No. 1 pen, white for No. 2 pen, blue for No. 3 pen—red, white, blue equals pens 1, 2 and 3, very easy to remember. Use the near-side leg for Conference ring with spiral above it, or below it if wished.

Thus when picking out birds from a flock you know if you want a pen 2 breed bird to catch one with a white spiral on near-side leg.

Later when mating up a pen or pens of birds, leave the pen colour ring on (for life) and ring the whole pen with one colour of spiral on the offside leg. Thus for the whole time on the farm you know how they are bred and which birds belong to the pen if they get mixed during out of breeding season time.

Often the question of age in young birds crops up—such as at Christmas time when picking out birds to keep or for stock, naturally the earlier birds are more likely to give you results in their first laying season, certainly better than late May hatched ones.

If you use the Conference rings and ring the birds in rotation, *i.e.,* use ring number in rotation starting with the smallest number, you of course know which are the oldest ones.　If you do not ring you can toe-mark on the small web which is found on the inside of the feet in waterfowl (see diagram).

Another useful marking, and a permanent mark, is to use " Wing Bands." These are very useful for some purposes.　I do not much care for them as they cannot be seen without catching the birds, as they are under the wing. Thus you cannot walk amongst the birds and know how they are bred, etc., as when using coloured spiral rings. Spiral rings are far better than flat rings—the flat rings seem to come off too easily.

There are quite a few other ways of marking but I think I have mentioned those most useful.

CHAPTER IX

CONTROL OF GEESE

I HAVE been questioned so many times as to what height of wire netting to use for geese to be kept in pens, etc., that it is perhaps best to write on the same.

I use 4-ft. netting with 2-inch mesh—just ordinary class of poultry netting. When erecting pens which are wanted for a number of years, I use light 3-inch mesh sheep netting, 42 inches high. In all cases I leave the stakes about 8 to 12 inches higher than the netting and run a wire on the tops of the stakes to which the top selvage of the netting is fastened. The heavy breeds never attempt to fly out—the lighter breeds will often do so, especially on windy days.

The quickest and best way to prevent a goose flying is to cut the feathers of one wing; by cutting the feathers I do not mean taking a huge pair of scissors and cutting nearly all the feathers away, thus spoiling the appearance and in many cases causing a gander to be very infertile. All that is required is to cut off about four inches from the ends of the first four or five long flight feathers of one wing; do this, and the birds will not fly over a fence; on the other hand they remain active and can nearly fly, yet not get above six inches off the ground.

Any gaps or holes in natural fences should, of course, be repaired and stopped. Geese are not difficult to control under ordinary circumstances, and even if they do go wandering they always return towards evening once they have become accustomed to their surroundings.

It is, of course, understood that many of the lighter

breeds, such as Chinese, Romans and the ornamental varieties, can generally fly when they are in the mind to do so, in which case one wing will have the flight feathers cut as mentioned earlier in this chapter. The cutting of the flights only lasts until the birds acquire new feathers, when the bird or birds must again be operated upon.

CONTROLLING GEESE.
" I use 4-ft. netting with 2-in. mesh . . . leave the stakes about 8 to 12 inches higher than the netting. . . ."

To those who wish the birds they rear to be permanent non-fliers, the best way is to pinion the goslings when hatched and " fluffed out." Here I would say that the pinioning of adult birds is very cruel and must cause much punishment to the bird operated upon and it is certainly done at the risk of losing the bird. The operation is carried out briefly as follows : Take a sharp knife and remove the piece or extremity of one wing. In a young bird it is very soft bone and gristle. What you are really doing is to remove the piece on which, later, the long flight feathers grow. Carry out this operation on one wing and your bird will never fly.

While my own farm is bounded by a river, and this, of course, is no obstacle to waterfowl, the only precaution I take against the stock wandering is to erect the four-foot wire netting on fairly substantial stakes, such as are seen in the photograph on the opposite page.

Geese are, however, past masters at discovering gaps, and, once having coming across such an outlet, they will follow one another through it like a flock of sheep. Take care, therefore, to keep the staked netting in good order.

If you have geese which can and do fly and which you have newly acquired and wish to settle on to a lake or water and yet later have them flying about your place as if in a wild state, cutting one wing generally settles them in. In the case of certain wild varieties of duck, and near the nesting season, I found the pulling of about six or eight flight feathers from one wing settled them in and later it was delightful to see them flying around. Again, it is to be understood that the birds were settled in suitable surroundings which had reasonable attractions for them and suitable for them to carry out their domestic duties.

I have gone through a few ways of controlling geese, but the best way is to have the birds happy and contented in their home and pasturage; under such circumstances they are no trouble or worry.

CHAPTER X

FATTENING

UNDER ordinary circumstances and with good class well-reared geese, kept in good store order, a month to five weeks good feeding before required should give a plump, fat bird. In the case of early goslings, or any goslings which are wanted for table purposes when just in full feather—in other words, used as a large Aylesbury duckling is—there should be no question of any finishing period as they should be well done all the time.

A " Michaelmas Goose," a green goose—meaning a bird killed in September-October, should be well fed and requires hardly any special foods.

In the case of the goose for Christmas we have a different proposition as the birds are required at a time of the year when most grass and natural foods are becoming very scarce. Thus most foods have to be provided for them, and for the last month it certainly pays to give them a special fattening menu.

The goose, fortunately, is not a faddy bird and will eat with relish much boiled vegetable matter, such as swedes, potatoes and turnips—these, in many districts and on many farms, are a reasonably cheap food, are on the spot and may be bought without any extra cost for carriage or cartage. There is, of course, the labour and expense in fuel to be considered and it rests with the owner to make his own calculations and costings. Home foods, plus labour, against bought foods and labour. Any boiling of foods takes place at the time of the year when most outdoor field work is out of the question and often an odd man could keep his eye on the copper of food.

All roots should be carefully washed and cleaned before boiling. All vegetable matter should be thoroughly boiled, in many cases faggots and odd wood could take the place of coal and thus save expense, here again it must be decided by the owner as to which is the cheaper fuel in the long run. A mash tub and masher should be provided in which to place the boiled vegetable matter, which should be thoroughly mashed up ready to dry off with meal.

Good meals are barley meal, maize meal, flaked maize and greaves (greaves provide the extra fat), all of which can be purchased at most good food stores. The whole is well mixed and fed in troughs, not a wet sloppy mixture and not too dry. Ample drinking water should be supplied.

In my opinion it is folly to shut geese up in a tiny place on the dark side with the idea of fattening them, generally the opposite occurs as they are given to fretting. The ideal place is a good fold yard with open-fronted shed. Failing this I would much prefer to let the birds have reasonable liberty on the ground on which they were reared, even allowing them access to water at all times. Certainly confine them to a smaller space if possible so that they do not go all over the place and take violent exercise.

If you do not use boiled vegetables, feed plenty of mash with lots of maize flakes, barley meal, greaves or a little fish meal, but on no account more than ten per cent. of the latter. The birds love grain and, when fattening, it is better to place it in troughs with water on it. The great thing to avoid is the use of expensive foods—make every use of easily procured food as suggested in the boiled vegetable matter, but avoid cheap grain.

A useful way is to boil whole maize; it is very fattening and the geese devour it eagerly. Make up your mind which birds you wish to fatten and once they are placed in a pen or in a certain place, on no account move them or take any odd birds away.

A man who yearly " fats " eight to ten thousand geese, passed the following remarks to me once : " Never sell or take out a bird or birds from a fattening flock—does much harm. I never starve the birds before killing, but do not feed the day they are to be killed and plucked. If I did starve them I should lose a lot of my profit."

If you have spare milk, skimmed or whole, buttermilk, etc., make use of it; it puts a finish to the birds.

Killing : The bird should be caught by its neck, the wings crossed on the back, if you do not care to do this, wrap a clean sack round the body to prevent the fluttering of wings. The neck should be dislocated by pulling.

A very simple way is to procure a round piece of wood, such as a broom handle, place on a level floor in front of you, your left foot on one end. Now place the bird's head under the wood with beak away from you, and the wood just at the point where the back of its skull joins on to the neck, once in this position put your right foot on the other side of the wood, now give one strong even pull. The round wood will go in behind the skull and sever or break the neck. Difficult to put on paper, but in action very quick, no pain, and sudden death. You are certain to have broken the bird's neck if the wood goes down to the ground, as it has severed the head from the neck except for the skin and feathers. There is no blood, no mess, yet space for blood to drain into when the bird is hung up.

In the sale of fat geese, whenever possible sell to private customers. Failing this get into touch with a good fish and poultry shop. Sell at so much per pound live weight and he does the rest, or sell at so much per pound rough plucked. Do not wait until the last minute before Christmas.

The goose is fast coming into its own again as a popular dish. We can help very much by providing a really first-class well-fed bird which will give every satisfaction.

CHAPTER XI

PREPARATION AND SHOWING OF GEESE

THERE is really little to be said on this subject. First of all it is essential that you have good birds before going to the expense of showing them; owing to their weight they cost a lot in carriage.

Having what you consider a good bird or birds which you wish to exhibit, the great thing is to have them really fit, fat and in good bloom and in perfect feather—if white birds, well washed. Of course, I mean naturally washed in water. The bird or birds should be caught, any private rings removed, bill and feet cleaned and just a little vaseline well rubbed in, placed in a roomy, well-bedded hamper and placed on rail to the show.

There is hardly ever any need to put birds in a training pen. I like to do so myself as the birds then get used to eating and drinking in the pens. In the case of Toulouse, being used to a show pen is an advantage, as a frightened, scared Toulouse will not drop the keel and show to perfection. Also the bird which sulks in the back of a pen does not stand so much chance as the one which stands up proudly in the middle and makes a show—asking to win.

In the case of white geese, especially if your place is wet and muddy, as it is difficult to catch a bird without getting it dirty and stained, the following is useful. Get the birds used to a shed or house. On the morning of dispatch, have the house well bedded with clean, bright straw, get the birds into the water to wash, later get them out and drive them into the house without causing them to fall or stamp on each other.   Close them in and leave for a time and later catch your bird, which should be unstained.

The show hampers should be well bedded with clean straw—my birds receive a good depth of straw on which I like to put some clean cut " chop."

EXHIBITION GEESE
The Toulouse (left) and Embden (right).   Both are well-known prize winners.

If you have good birds, make a point of showing one or two each year at such as the Dairy Show (for young birds).   Visit the show and see results, talk to other breeders and get to know where and how your birds fail or excel.

I said show your best, meaning, of course, what you consider your best. It is essential that you have some idea of the chief points from a show point of view of the variety or varieties you are showing.

To give a rough idea of what I mean we will first imagine the reader has what he or she considers a really good lot of Embdens.   In picking a bird to show you certainly want to send off a good, big bird, but remember size is certainly a big asset, yet a medium-sized, good-shaped bird, perfectly free from keel and gullet and dead white in colour, shown in perfect condition will, under most judges, beat a very big bird showing a lot of gullet, or a gander shown dirty or slack in feather. Condition counts every time, and the exhibitor who puts his bird down in perfect bloom receives due reward. A good big one will generally beat a good little one.

Show some of your best birds each season; better still go to some of the best shows and try to meet other breeders and exhibitors. Whenever possible have a chat with the judge, but don't grumble at his awards!

White varieties are certainly the most difficult to show in perfect bloom and free from stains. Under certain circumstances it is often necessary to wash geese for show. I have seen a number of washed birds, but they never compare with the unwashed ones as they lack the lovely bloom and sheen of a naturally washed waterfowl. It is certainly far better to make an effort and hand-wash a bird than to show it in a dirty, stained condition. Do your best and show your best, but do give it every chance to win by paying some attention to its preparation.

CHAPTER XII

THE EMBDEN

IN the Embden we have without a shadow of a doubt the most useful and sound variety of goose. It answers all useful purposes and gives exceptionally good results in most situations, and proves hardy, vigorous, prolific, and generally gives excellent fertility combined with good hatchability and rearability. At the present time breeders have got the Embden to a fine pitch of perfection for all general purposes.

A few of its good features are snow-white colour, thus giving feathers of first-class quality, flesh of extra good colour and texture, plenty of meat on a good, plump breast. Matures early, and is thus useful for Michaelmas gosling. The female is a good sitter and makes an extremely good mother when used. Lays a good number of eggs, usually from 20 to 30 in first laying, followed by 10-12 in second batch, of course depending on strain.

The gander is most useful in that he can be used to cross on to other varieties, such as the Toulouse or Roman. Used as a sire with any variety of goose it will be found that the table properties of the progeny will be vastly improved, especially so when used on to dark-fleshed varieties.

Without a doubt the Embden is a direct descendant of the common white goose, and it can reasonably be proved that white geese were in existence in 50 B.C. and even before that period. I think most nations will recognize the perfecting of this beautiful bird by English breeders. Granted that Germany is the home of the Embden, and that we got them in the first instance from that country, taking their name from the town of Embden, yet since they came here they have improved wonderfully in type and size, so much so that they are now in front of the German Embden. Germany, in fact, has bought from us to improve her stock for many years.

The Embden is often described as " swanlike." It also mentions " swanlike neck " in the breed Standard. They are certainly not unlike a swan at a quick casual glance, but to my eye and mind you spoil the Embden when you breed it on the lines of a swan.

The Embden should be a big bird with a large compact body, not too long in the neck, and the same must not curve as does the swan's neck. Certainly a neck of good length, but not long and curved. Again, we do not want a long thin body as in the swan, but a body of good length combined with lots of width and depth.

A careful study of the photographs in this book will give the reader a good idea of what is wanted in an Embden—a bird with a perfectly clean outline, medium length of feather and free from signs of keel and gullet. It may be best to try to explain clearly the terms " keely " and " gullety."

The gullet or pouch is the piece of skin with feathers on the throat as in the Toulouse. In this latter breed it is looked for, but it is a very bad fault in an exhibition Embden, a point to be carefully watched when mating up a breeding pen. A striking illustration of this point is seen in the photograph of the Toulouse and Embden geese on page 52.

Many of the biggest Embdens, which have at some time had a cross of Toulouse to improve size, are inclined to show coarseness in gullet and keel, points which should not be

A BATCH OF EMBDEN GEESE
The Embden is "a bird with a perfectly clean outline, medium length of feather and free from signs of keel or gullet."

allowed in any birds from which you hope to breed winning stock.

By "keely" I mean that the breast and underline of body up to the paunch is as in a boat. A good Toulouse shows plenty of keel, but in Embdens any suspicion of this keel must be avoided. What is wanted is a perfectly rounded breast, apple-shaped when viewed from the front and a bold clean-rounded outline when the bird is viewed from the side. The best type bird will, of course, show a little sign of this fault when slack in feather, or if in poor thin condition, so due allowance must be made at certain seasons of the year.

As a breed the Embden is extremely hardy and is a good forager. Goslings are easy to rear, mature quickly, and once they have some feathers on their backs require no housing. I never house any Embdens, winter or summer, and as for foxes harming them, well, I have yet to hear of the fox that would tackle one of my ganders, or one of his wives whilst he was near.

There are just one or two points which often cause the beginner some anxiety with this variety. First, the goslings do not generally come yellow in colour, many come with grey colour on the back down. I often find the darker goslings become the best white when in adult plumage, and are usually free from that objectionable sappy white colour. The other is that some will come with a few odd dark or grey feathers on the rump (often denoting a female gosling). These generally clear when the bird gets its adult plumage.

In some strains we get a small percentage of birds that feather-up white-flecked with light grey and some grey in flights. This is certainly a throw-back to Toulouse blood in past breeding. These birds are useless as show specimens. On the other hand, they are often the biggest and weightiest, thus they are excellent for general-purpose breeders.

A white variety of waterfowl never looks at its best without swimming or, at least, washing water. Under ideal conditions, with plenty of water and grass, there is no better looking or more graceful bird than the Embden.

In their first laying season Embdens are not always a great success, especially any geese which are late hatched. They may lay lots of eggs which prove most fertile, but generally the goslings have difficulty in getting out of the shell. On the other hand, once you have a good pen they should give you good results for at least ten to twelve years, even more. I have a goose that laid over twenty eggs in her fourteenth year, and from which I hatched sixteen goslings. The gander will give very good results mated with three geese. If an old gander, and fertility was not good, it might be best to reduce the geese to two.

I have had 99 per cent. fertile eggs from four first-season geese mated to a first-season gander. What is more, they are all giant birds. Many of the eggs hatched out well and the resulting goslings were quite rearable, of course requiring a little more attention in the early stages. With correct breeding stock in this variety it is to be expected that one can average 20 lbs. dead weight at Christmas on the crop, taking it that the goslings were given a good start in life and were fat when killed.

POULTRY CLUB STANDARD OF PERFECTION FOR EMBDEN GEESE

GENERAL CHARACTERISTICS

HEAD, long and straight.

BILL, fairly short, stout at the base.

EYES, bold.

NECK, long and swanlike, the throat uniform with the under mandible and neck, *i.e.*, without a gullet.

BODY, broad, thick and well-rounded; round breast with very little, if any, indication of keel; broad shoulders and stern; long, straight back and deep paunch; large and strong wings; close tail, carried well out.

LEGS, fairly short; large and strong shanks; straight toes connected by web.

CARRIAGE, upright and defiant.

PLUMAGE, hard and tight.

WEIGHT, gander, 30 lbs. to 34 lbs.; goose, 20 lbs. to 22 lbs.

COLOUR: Bill, orange; eyes, light blue; legs and feet, bright orange; plumage, pure glossy white.

Type (breast 20, head 12, general carriage 12, neck 10)						54
Size	...	...	...	...	...	20
Colour	...	...	...	...	...	10
Condition	...	...	...	...	...	10
Legs and feet	...	...	...	...	...	6
						100

SERIOUS DEFECTS: Plumage other than white; any deformity.

CHAPTER XIII

THE TOULOUSE

THE Toulouse is in every way a fine handsome bird, beautiful in colour and outline, and a variety which is most useful to the fancier who wishes to try his skill in breeding for special points, yet have a sale for the birds either at a big price for his best efforts, or a purely commercial price for those which do not quite come up to show perfection.

This variety is an absolute contrast to the Embden both in type and colour—lots of keel and gullet are wanted; also big size, long body, with as much width and depth of body as the breeder can procure.

To breed really first-class exhibition specimens is not easy. It takes many years to build up a strain which will give a number of birds of good type combined with the correct soft shade of grey.

This variety is of French origin, taking its name from the town of Toulouse, but here again, as in the Embden, English Toulouse have been bred to such perfection that they are now, as a whole, far in advance of the average French birds. I make this statement without fear of contradiction

¯as I have had many letters from French breeders who year after year import English birds.

This variety is not nearly so active as the Embden, and a gander generally does best with two geese. There are, of course, exceptions, and many ganders safely take three geese. It is up to the owner to use his judgment. Certainly an old gander is best with two geese, and in cases when wishing to breed in special features, such as keel, gullet, or even colour, I like pair matings.

The Toulouse goose on the average is a better layer than the Embden. One can safely say it is the best layer of all heavy breed geese. I have had 60 eggs in one season from a three-year-old goose, and no signs of broodiness shown. Others have laid 40-45 eggs. Many strains show no signs of broodiness, but those birds that do are quickly and easily broken. Certainly they should not be trusted with eggs. If you do give them a nest they generally throw up the job!

The standard of perfection and photographs shown will give a good idea of what is required in a good typical Toulouse, without going into details here.

In choosing your breeding stock pick birds as large as possible, combined with a good soft shade of grey. Beware of the poor common dark, dead shade of grey; once get it in your strain and it crops up generation after generation, generally in your best bird as far as size and type are concerned.

Choose birds of extra good length of body, as low as possible to the ground, and with this see that you have plenty of width of body. Many which appear extra good birds when viewed sideways appear perfect, but when seen from above and going away they are spoilt by extreme narrowness of body. I am very keen on this latter point as in my opinion a goose which is narrow across the top has no lung space. Choose birds with the good points above mentioned. combined with plenty of keel and gullet—the longer and deeper the keel the better.

The way and time to inspect a Toulouse is when the bird is standing in a natural position; obviously the bird will not show its keel line to perfection if frightened or rushed into a corner. Do this and the bird will take up an erect stance, or as near erect as possible. The Toulouse will generally show well when walking about naturally and unscared.

On no account throw out a good bird because it lacks size. Given good type in a well-bred Toulouse, one can soon get size with careful management and judicious mating; on the other hand, it is not much good breeding from a big bird lacking in colour and poor in type.

The goose of this variety is much used mated with an Embden gander, and such is certainly a grand cross in many ways. A cross gives more stamina and hardiness. The Toulouse goose is nearly non-broody, and lays a larger number of eggs than does the Embden, and by using the more active Embden male more geese can go to the pen, often four geese.

Fertility is generally better when using the Embden gander and the resulting progeny are easy to rear, as it is a direct cross. The goslings mature quickly, and make grand Christmas or Michaelmas birds. What is better still, the flesh of the Toulouse is improved by using the Embden, as the goslings come white, grey and white splashed. The lighter-coloured birds pluck out much better.

With the cross not many come with distinct and clearly defined grey and white colour markings. Hundreds of birds are sold each year as Embden x Toulouse—" grey and whites." They may be full of Embden blood, but on one side they are most certainly descendants of the good old English " grey-backs."

Thus one can safely say the Toulouse is a most interesting variety to keep and breed; useful from a utility point of view, interesting and fascinating to breed to the standard of perfection for exhibition.

If one has the space and accommodation it is nice to keep a pen of both Toulouse and Embdens, as one shows the other off, and gives a beautiful contrast of what the breeder can do in colour and type; especially so if you have in mind that all geese are descendants of the wild Grey Lag Goose.

A SPLENDID PEN OF TOULOUSE GEESE.
Of this popular breed Mr. Appleyard says : " It provides an absolute contrast to the Embden both in type and colour—lots of keel and gullet are wanted."

It is essential that the Toulouse has water for mating purposes. If you have not a natural pond or stream it would be wise to make a bathing pool as suggested elsewhere in this book. It will help to give you the desired results.

The Toulouse generally looks bigger and more massive than the Embden. Such is not really the case, it is simply that the Toulouse carries more and looser feathers, also the keel and gullet add to the impression of greater size.

C

To anyone interested in breeding birds for special type and colour, I would certainly say try this variety, it will give you plenty of scope to show your skill, plus the fact that you have a useful bird for utility purposes. Here let me say to the farmer or breeder who only requires birds for table purposes that the keel and gullet is of no use for table purposes; in fact, a bird without it is, if anything, better. It is thus not necessary to pay special exhibition prices, as birds which fail in these points will give just as good results as far as utility properties go.

The photograph opening this chapter depicts a really first-class exhibition specimen, clearly showing the gullet or pouch (the French term it " bavette "), the hanging part under the throat. The keel is shown but can only be seen side view and in outline. A good keel should come to a point when viewed from the front, just the exact opposite to the Embden which, as explained, should be rounded, or what is termed " apple-shaped," when viewed from the front.

POULTRY CLUB STANDARD OF PERFECTION FOR TOULOUSE GEESE

GENERAL CHARACTERISTICS

HEAD, strong and massive.

BILL, strong, fairly short and well set in a uniform sweep or nearly so, from the point of the bill to back of the skull.

EYES, full.

NECK, long and thick, the throat well gulleted.

BODY, long, broad and deep; prominent breast, deep and full, the keel straight from stern to paunch, increasing in width to the stern and forming a straight underline; broad shoulders.

BACK, slightly curved from the neck to the tail; large and strong wings somewhat short tail carried high and well-spread; paunch and stern heavy and wide, with a full rising sweep to the tail.

C*

Legs, short; shanks, stout and strong-boned; straight toes connected by web.

Carriage, somewhat horizontal, not as upright in front as the Embden, and thick-set.

Plumage, full; somewhat soft.

Weight—Gander, 28 lbs. to 30 lbs.; goose, 20 lbs. to 22 lbs.

Colour : Bill, legs and feet, orange; eyes, dark brown or hazel; plumage—neck, dark grey; breast and keel, rather light grey shading to thighs; back, wings and thighs, dark steel grey, each feather laced with an almost white edging, the flights without white; stern, paunch and tail, white; the tail with a broad band of grey across the centre.

SCALE OF POINTS

Type (head and throat 15, breast and keel 10, tail, stern and paunch 10, neck 5, general carriage 15)	55
Size	20
Colour and markings	10
Condition	10
Legs and feet	5
	100

Serious Defects : Patches of black or white among the grey plumage; slipped or cut wing; any deformity.

CHAPTER XIV

THE ROMAN

THIS variety is well worth consideration by those who do not require a big heavy goose. When dead and dressed it certainly pleases those who like goose, yet do not want an 18-20 lb. bird on the table. The Roman is all quality, good coloured flesh, very light in bone without a lot of offal. The flesh is delicious eating, and with ordinary feeding and without any special fattening the Roman generally carries lots of meat.

Pure white plumage, delightful outline, orange pinky bill, legs and feet. Very active, alert, most fertile, eggs hatch really well and the goslings are very easy to rear. The geese are extra good layers of medium-sized eggs, some strains lay up to 45-65 eggs in a season. Those I have sleep in the open, are never housed and even when given a shelter preferred to live out on the snow and were quite happy and contented. In my opinion, this variety should be kept small, combined with good all-round utility qualities, *i.e.,* eggs, flesh and quick maturity.

In outline they should be less rounded in body than the Embden; rather long cast in body, short on the leg, not too long in neck. Show a little longer length of tail, heads very refined in character, especially in the goose, short, sleek, close-packed feathering. In the geese you may get a few come with flecks of grey on the rump, but are covered by the flight feathers, thus they do not show and may be quickly bred out with a little care in mating up the pens.

Although not so popular as some other varieties there are many nice-sized flocks of Roman geese in Britain.

The gander is very active and will generally take up to four geese, often five. To my idea Romans prove most useful to those who have a small grass paddock, or any spare grass. A breeding pen costs little to feed if a plentiful supply of grass is available. Goslings may be used as if they were Aylesbury ducklings, *i.e.,* killed when just in first full feather, later, when in second feather, and then onwards up to and for Christmas. Any late eggs should certainly be put down and any resulting goslings will give birds to kill in the new year.

For the above-mentioned reasons, I consider I am right in saying that Romans are most useful for any private house, or country hotel, in fact for all who would care to produce a delightful and delicious dish for their own table at a little cost.

For those who wish to run only a few geese the Roman fits in admirably if a small paddock can be turned over to a set of four or five. They are really excellent grazers and will keep the grass trimmed to a nicety.

A sound plan is to divide the paddock into two halves, so that while one half is in use the other is resting and recovering from the grazing. By careful use of each half— the length of time either one is in use depends a great deal on the weather and the season—the geese always have fresh grass on which to run.

At certain times of the year, such as when youngsters are being reared, both halves will be in use. But, as I have already said, careful planning and management will keep the turf in good order.

It is essential that the Roman is bred and kept as a small breed combined with the desired utility points. The photographs will give the reader a good idea of the Roman. To those who like goose, " Michaelmas goose," or "Aylesbury duckling," I would say try a Roman gosling cooked and served with the accessories which generally go with duckling and you will enjoy a delicious dish.

CHAPTER XV

THE CHINESE OR AFRICAN

I HAVE placed these two under one heading as it is my firm opinion, after much thought and care, that the African goose is chiefly Chinese blood with a cross of grey goose in it. This blood has given the extra depth of body and the slight gullet. Apart from this extra size and gullet they are identical to the Chinese. A few show less of the Chinese colour— are of a darker, duller shade.

The Chinese goose (*Anser cygnoides*) has been classed with the swan; it very much resembles the swan in outline, especially in neck and head, and knobbed bill. Another thing is that the neck feathers are smooth and do not curl as in other geese. It breeds freely with any goose, thus is of the goose tribe, as the progeny from, say, Chinese gander and Embden goose are fertile.

The original Chinese variety ranges all over China, in Siberia and India. In size it is a small bird, most hardy, never sick or sorry, always on the alert. In appearance it is beautiful and elegant, is the most ornamental of all the domesticated varieties of geese, and it goes and stays on water—absolutely loves it, spending much time diving and splashing about.

On the other hand, it is a variety perfectly fertile and con-

A " SET " OF CHINESE GEESE.
" The Chinese goose very much resembles the Swan in outline, especially in neck and head and knobbed bill."

tented, and always appearing clean without any swimming water.

The Chinese goose is most prolific, often laying up to 50 and 60 eggs in the year. Instances of 100 eggs per year are known. This fecundity to my mind is most useful as the geese may be mated with a medium-sized active Embden gander (preferably bred from large stock) or a good-sized Roman gander. Such a cross is most helpful, as it gives lots of eggs which can be turned into goslings. The Embden or Roman blood improves the size and flesh, giving quite a useful bird. I have tried a number of experiments in crossing them, and when the first edition of this book was in course of preparation I was trying out an Embden-Chinese cross.

This cross gave offspring of a delightful soft silver-grey shade and, while they followed close to the Chinese type, they showed clearly in their deeper and wider bodies the stamp of the Embden.

I have continued using this cross, and I do not hesitate to recommend it to those who want to bring in a fast-growing goose of true utility type.

This crossing of our better breeds on to Chinese is most useful, especially for those who live in exposed situations with poor grass and herbage. The Chinese gander may be safely mated with four geese, even five. They are most fertile, eggs hatch well and the goslings are hardy, quaint little fellows, very independent and easy to rear.

The flesh is dark and with a flavour all its own. Many prefer it to other goose flesh. It certainly is useful for the delicate person who does not care for too much fat.

The Chinese is occasionally shown, but there is no standard of perfection as far as I know. The plumage is charming; very tight, short feathering, always neat and tidy, with a clear, clean-cut outline. In colour they are many shades of brown with whitish under-parts, with a darker brown stripe running down the back of the neck. There is a white variety of this breed. Some were exhibited at the World's Poultry Congress in England, and were very beautiful indeed. A picture of these birds appears in the chapter on " Ornamental Geese."

Of all breeds the Chinese are wonderful watch-dogs. They speak the moment a stranger puts foot on the place, and keep speaking. On the other hand, they are most intelligent and friendly with their attendant. The goose is a good sitter and mother. Allowed to carry on naturally she will make quite a good job of it and will bring off and rear her own young most successfully; the gander makes a very kind and good father.

To those who want beauty combined with utility, the Chinese is certainly well worth consideration. The eggs are medium-sized, mild in flavour and excellent for cooking purposes—thus eggs not required for hatching prove most useful. The Chinese goose generally lays a number of eggs in September and October. This is something that happens only once in a " blue moon " with other breeds of geese.

CHAPTER XVI

USEFUL UTILITY VARIETIES

THERE are many varieties of geese, very varied in type and colour which would well pay to take up—pay both in interest and in money. The beauty of the goose is that one can sell at a remunerative price any birds not required to keep on as future breeders.

English Geese.—I am placing under this heading a number of varieties, many of which are hardly ever bred specially for type, colour, etc., yet, as I say, would well pay if taken up and bred to a standard. You would, of course, have to make your own standard as there is no written standard.

"*Grey - backs.*"—I use the term "grey-backs" as it is much used to denote a very cobby, thick-set type of goose which has a sheet of grey on the back. With this one generally sees grey feathers on the thighs and some grey on the head. By careful choice one could quickly get them evenly marked and the even marking on a snow-white ground would look

73

really charming. In addition one could breed for size combined with fecundity, etc.

Greys.—Here we have another charming goose in many shades of grey, from dark, nearly black, grey, right through the greys to a near lavender shade of grey. Many of these " English greys " are really wonderful layers. I have heard of one which laid 80 eggs—eggs which hatched, too.

Crested.—Many geese carry crests, or a tuft of feathers on the head. I know of no case in which they have been bred especially for type, size and shape of crest. Many old paintings of farm scenes depict geese with crests, in some cases with large and excellently shaped crests—birds of all colours, greys, whites, greys and whites and splashed-coloured birds.

Buffs.—Yes, buffs! There are buff geese! I have quite a few, really charming shaped birds, medium-sized, very round, compact bodies, small neat heads with rather short, thick bills. As yet I have not got them to perfection, nor have I made up my mind as to the exact shade wished for, whether to try to breed them an even soft shade of buff throughout (*i.e.,* buff instead of white), or to breed them buff colour yet marked just as a grey goose with buff in place of grey. At present I, and those who have seen them, prefer the latter colouring, that is to say, buff-marked. They have proved hardy, good layers, fertile, pluck out a good colour and are very plump.

Blues.—Is it possible to breed a blue goose? Time will

tell.   However, it is too early for me to say, but I do think
and have good reason for saying there will be quite good
coloured blue geese before we are much older.

The Sebastopol.—This variety is now rare, and should
really be placed under the heading of ornamental geese;
yet it is just as good as other varieties from a table point of
view.   Medium-sized, snow white, with lots of long
feathers on its back and long feathers hanging down, often

A trio of Sebastopol Geese. The flowing mane of feathers, more pronounced
in some than in others, makes this variety most attractive.

touching the ground; hardy, free breeder, very fertile, very
quaint and out of the ordinary, and most certainly arouses
the interest of all who see them.   Mine have caused many
who are not much interested in geese to stop and " sit up
and take notice."

For some unknown reason not much interest is taken in
making new colours and shapes for geese.

Bantam Geese.— What about breeding bantam geese? At the mere mention of bantam geese many would consider the speaker mad, or, anyway, on the way to madness.

What would there be against a bantam Embden, which would nearly live on grass, require no housing (other than protection from foxes and dogs, etc.), and at 10-11 weeks of age give a delicious 5 to 6 lbs. dish? Or, later, a 6 or 7 lbs. goose!

I think they would go well and be most popular. Anyhow, I have some, decently reared, pure white, blue eyes, passable type which, at five and six months of age, do not weigh more than 5 to 6 lbs! Compare this with a six months old 20-22 lbs. Embden, and it is not too bad. I have *none* for sale, and do not write about them with the idea of making a sale.

One good feature about geese is that you can carry out much that is of interest from a breeder's point of view and yet not lose money. Generally speaking, what is of no use to the breeder for the future can be easily disposed of as gosling or " Christmas goose." Thus, I see no reason why many new colours and shapes of geese should not be added to those we already have.

CHAPTER XVII

ORNAMENTAL GEESE

THERE are many other varieties suitable for the hobbyist, such as : Emperor Geese, Blue-winged Geese, Ross's Snow Geese, Blue Snow Geese, Magellan Geese, Ceveoposi's Geese, Orinoco Geese, Red-headed Geese, Maned Geese, Falkland Island Geese, Red-breasted Geese, Ashy-headed Geese, Bar-headed Geese, and others, all of which do quite well in captivity and many of them will breed if in a really good situation. Many of these birds can be purchased in England from those who specialize in fancy waterfowl. They are catalogued at prices from three guineas to seventy-eight pounds per pair. Many are reasonable in price, and to those who have protected quiet waters they are certainly well worth while from the ornamental point of view and for the great interest they give.

I give a brief description of the commoner varieties often kept on private waters :

The Grey Lag.—This variety is said to be the ancestor of all our domesicated geese—thus, it is responsible for a lot. It will cross with any of its descendants (size permitting) and give fertile progeny. In type it is very narrow and shallow in body with good length of leg, quite unlike its descendant, the common Grey English Goose. In colour, brownish-grey with lavender-grey shoulders. Bill, flesh-colour to orange-yellow; legs and feet, flesh-colour; claws, black; iris, brown.

The Grey Lag still breeds in Scotland and will breed in captivity. Lays from six to eight eggs. It is certainly of interest to keep and breed in captivity or in a natural state, pinioned on lakes or meres.

The Canada Goose.—Charming variety, of beautiful outline, hardy and free breeders. Many cases of crossing have been made with other varieties. One season I had

A pair of Canadian Geese. Strong fliers, they have to be pinioned when young if they are to remain at home.

a Canadian gander which mated with a Chinese goose and gave 90 per cent. of goslings from eggs used.

The Canada goose can fly very strongly, and in many parts of England now lives and breeds in what may be termed a wild state—generally living on one sheet of water, but paying visits to other waters. At Holkham, in Norfolk, the seat of Lord Leicester, one may see as many as a hundred birds at certain seasons of the year. They spend much of their time on the water. They are of good size, very good table birds, with a gamey flavour of their own.

When commencing with Canada Geese it is best to get a pinioned pair, or rear young pinioned birds which will have to remain on the spot. Their progeny could go unpinioned,

White Chinese Geese. First of these were seen at the World's Poultry Congress in London in 1930.

as there is a certain charm in having birds which can and do fly about, yet return to their homes. Some may get lost or even shot, even then those which remain are most interesting to watch in flight and movement.

The photograph gives a really good idea of this charming and interesting variety—dark brown in body colour, the bill and legs black, black head and neck, the head set off with a white crescent across the throat. Very short, compact, crisp feathering—a real " dandy " in appearance.

The goose makes her nest near the edge of water, generally a large structure of rushes and reeds, lined with down. The eggs, generally six to eight in number, are laid early in April.

The Barnacle Goose.—The Barnacle Goose is small, about half the size of the Canada. In colour it is very striking, with its black crown and neck and creamy-white face. The body is palish grey and the feathers have a black bar before the usual pale tip. This double working gives quite a striking and gay appearance. The bill and legs are black, under-parts of body white, as are the sides of the rump and upper tail coverts.

This species nests in Greenland, Spitzbergen and in the Arctic; it is quite common in winter in Northern European seas. It is more frequent in the Hebrides than is the Brent Goose. It is plentiful in the North of Ireland and on the West Coast of Scotland during winter months, but is said to be becoming less each year.

It is said to breed well in captivity, making quite a success in rearing its young. It has been known to mate and breed with the Canada Goose and with the White-fronted Goose in captivity.

CHAPTER XVIII

AILMENTS AND DISEASES

SPEAKING generally, it can be safely said that geese do not suffer much from many of the dreaded diseases as do poultry. Geese are very strong, hardy birds, spending much of their life in the open air and generally in fairly natural surroundings and living on natural, non-forcing foods.

Geese are seldom ill; when they do become ill they are soon better or soon dead! Of all the breeds I have kept the Goose is the bird I have found to suffer least from diseases. I am not a scientist or veterinary surgeon and can only give a few symptoms and cures as I know them. If you have any trouble it is best to act quickly by sending a dead bird for *post-mortem* examination.

PNEUMONIA : It is seldom we find an adult goose suffering from this complaint. With goslings it is quite the opposite, as hundreds die each season, either from inflammatory condition of the lungs or from simple congestion due to actual pneumonia. In the case of goslings the trouble occurs through faulty, ill-ventilated housing, overcrowding and heating. The prevention is to see they are properly housed, with plenty of fresh air and a dry, well-bedded floor.

LEG WEAKNESS : From time to time one is troubled with this complaint in young goslings. From the age of four to twelve weeks is generally the period when one has a few birds go off their legs. It will be found in most cases the sufferer is one of the biggest and fattest goslings. The answer is in the rearing—lack of some essential vitamins in the foods fed, a bad, damp season, lack of sunshine, or feeding too forcing foods, which only go to form soft bone. Much can be done to prevent this trouble by adding cod-liver oil, minerals, yeast, etc., to the foods fed.

Each season a few of the very biggest and best " go off their legs." Some recover, some die. To make certain, mark the birds, sell at Christmas and on no account breed from any bird which recovers.

Not for one moment do I consider any cases of leg weakness due to the floor of the houses or to damp ground—although I would always have coop or house floor well bedded with litter.

In some cases this trouble may be partly hereditary, and it is best to let any bird which shows the slightest signs of weakness be marked and sold for table purposes.

BOWEL TROUBLE AND DIARRHŒA : In England we are not as yet troubled with any very dreadful bowel diseases. Abroad, where huge numbers of geese and ducks are kept together, they have cholera—a dreadful disease. I believe cases have occurred in England.

If you are so unfortunate as to have a lot of birds going light, moping about, suffering from very bad diarrhœa, the very best thing is to isolate all suspicious-looking birds and to confine all the flock to one place. Send some bodies away for *post-mortem* and to have a proper bacteriological examination made.

In certain countries where the disease is very prevalent, vaccines and serums have been prepared and tried with good success. Burn all dead carcases at once, destroy any sick birds and burn, disinfect the ground with quick-lime; any houses, etc., should be well disinfected.

If you have birds going light and showing signs of diarrhœa do not at once jump to the conclusion that your birds have cholera. They may have eaten some stale food or some weeds, etc., or at certain seasons of the year too many fallen apples or fruit. Worse still, they may have worms.

Worms: Unfortunately, not much is yet known about internal parasites in geese. We do know that worms have been found in geese. If you are feeding your birds well and they devour lots of food which does not appear to benefit them much, the best way is to suspect worms and, if necessary, kill the worst bird to send for *post-mortem* examination by a skilled man. If worms are there in any quantity the birds should be dosed with worm cures, which can now be had in pill or capsule form.

Colds: Occasionally geese suffer from cold—but it is very rare. By colds I do not mean roup. I have never yet come across roup in a duck or a goose.

♦ Geese, especially if ill-nourished and when moulting, will sometimes have slight nasal catarrh and show dampness around the eggs or frothy, white bubbles. In such cases (generally in old-aged birds or in ill-reared young birds when getting their second feather) the simplest and best remedy is to feed well on nourishing, soft mashes, to which add cod-liver oil. Get a suitable container (a small bucket or large tin), fill with warm water and make the colour of port wine by adding permanganate of potash crystals. Now take the bird firmly and immerse its head right under and keep under until the first sign of a struggle. Dip each day until better.

Staggers (Cerebro-Spinal Meningitis): Not much is known about this disease. It is certainly most disheartening, but, fortunately, not common. It can be briefly described as follows: Occurs at all ages, from a few days up to about 10-11 weeks of age, and the symptoms are: The affected birds walks about in dazed manner, often with the head on one side and held high, staggering about and going in circles. The bird may partly recover, but generally has another attack and quickly succumbs.

I will be frank and say I know of no cure. One thing is certain, it must be a germ, and a germ which can do its deadly work best when the earth is moist and warm.

Personally, I have had no trouble during hot, dry weather—always during moist, hot periods. Fortunately goslings ar not so susceptible to this complaint as ducklings. Imm diately a bird is noticed or even suspected remove at onc and look for any more likely or suspicious ones. The clean out the house, and if rearing on open field in wir compounds, move the whole to fresh, dry ground. Gosling at liberty do not appear to suffer so much from this disease.

OVARIAN TROUBLES: In geese we are not caused muc worry with ovarian disorders; from hundreds of layin geese I have had very few bothered. They were as follows Difficulty with a first-season goose in passing a big, doubl yolked egg. I simply took some lambing oils (olive oil wit a little disinfectant in it) and injected some, using my littl finger; she passed the egg all right. The other was a cas of eversion of the lower part of the oviduct. The goose wa restricted in comfortable quarters, the parts washed with mild disinfectant and the whole carefully returned and wa successful. In another case the parts would not remain i position and the bird was put out of her misery.

The Goose does not put up records of 250-300 egg per year, which is one of the reasons why she is not trouble with ovarian disorders.

LIVER DISEASES: Very rare but occasional cases occur i old birds as the result of gross feeding on very rich food If you suspect liver disorders, catch and examine the birds give a dose of Epsom salts and reduce rich foods.

CROP BINDING: An ailment from which geese seldon suffer. When a case occurs it is generally in an old bird Giving sour milk and kneading the crop can be tried, ofte with success. If taken early, it is possible to break up th lump of coarse grasses, etc. In advanced cases the onl cure is a surgical operation, which can be done if the bir is of sufficient value.

The operation briefly is as follows: Make an incisio and empty the crop—it is best to get some experienced perso to carry out this operation. The crop is emptied, washed out with mild disinfectant and then stitched up again, after which the patient is carefully looked after. Properly carried out quite 98 per cent. of such operations prove successful.

"Slipped Wing" or "Rough Wing": By "slipped wing" we mean that the flight feathers, generally in only one wing, do not lie in place properly packed. They stick out at right-angles to the body in the worst cases. There is no cure in bad cases. If a young gosling just in feather, the tip of wing to which the feathers are attached may be amputated, but it is best to clip the feathers off as short as possible and let the bird go for table purposes.

In the case of very special adult birds which show a failing such as five or six flight feathers hanging down a little, this can be put right when the feathers are growing by the use of tape. But it is hardly ever worth the trouble.

Never breed from a bird showing signs of this " rough wing." It is certainly hereditary and by using such birds a percentage of the progeny will come like it.

Tongue Down : Again rare, generally occurring in old birds of advanced age. Between and under the bottom mandible is a " floor " of soft, thin, leathery skin. This skin becomes enlarged and sags downwards, food, grit, dirt and grass get into it. You empty it time after time. The tongue drops into the cavity and thus the bird is in a difficult position—as fast as you remove the tongue it goes back.

In an advanced stage I know of no permanent cure, surgical or otherwise. If you act immediately you see the least suspicion of the trouble commencing, a cure may be affected by the use of a blister. Ask your veterinary surgeon for a mild iodine blister and instructions how to use it. Apply from time to time on the part. The idea of the blister is, of course, to form the thin skin into a hard, thick, shrunk, calloused skin and thus do away with any looseness. You must act quickly, otherwise it is useless.

LAMENESS, SWOLLEN FOOT : In hot, dry times, especially when geese have no swimming water, some birds get swellings and go very lame—generally it is a form of an enlarged corn. Get the lame bird up and place in a well-bedded house. Make a hot bread poultice, place this on the enlargement—in other words, get plenty of the poultice on to and around the swelling—and bind securely into position. Poultice a number of times until you have the affected part perfectly soft; now take a penknife and gently, but firmly, remove the core. If there should be any pus wash it out. Bathe the foot with disinfectant and cover the whole -foot to keep out dirt. It will quickly get in order and once clean and dry liberate the bird and return to the flock.

———

A companion book to " Geese " is " Ducks," by the same author. Dealing with all varieties, from the commercial as well as the show point of view, " Ducks " is obtainable through any newsagent or bookseller at 5/- or by post 5/4, from Poultry World Ltd., Dorset House Stamford Street, London, S.E.1.

Ground Insects for Grand Geese

THE susceptibility of poultry to such a wide variety of ailments has, for generations, given rise to much theorizing by experts. Many have held the opinion that physical tone has been lost largely by in-breeding, and that, with modern production methods, heavy losses amongst young stock must continue.

Insects an Essential

Over a quarter of a century ago, this problem of heavy mortality amongst "barnyard birds" was engaging a Manchester Chemist. An early query which this investigator put to himself was : " How does the diet of the domesticated bird—hen, duck or goose—differ from that of its ancestor —the Indian Jungle fowl ?

A consideration of the wild fowl's home showed that she had an unrestricted access to the insect life with which the Indian Jungle teems. Obviously, insect life in this unlimited quantity was denied to the domestic bird, living in an alien land, under " civilized " conditions.

Insects, it was realised, constitute a rich source of supply of certain tonic substances, including phosphorus, in an easily assimilable form. So concluded our investigator, this same natural stimulant that gave vigour to the jungle fowl, should bring the same health benefits to her cultivated descendants. Then followed experiments by the Karswood Company in which they succeeded in producing an article called Karswood Poultry Spice containing ground insects in a correctly-proportioned quantity.

That was over a quarter of a century ago. To-day, a vast multitude of poultry-keepers—breeding every species of the domesticated bird—are proving that the Manchester Chemist's conclusions were right up to the hilt.

Food Assimilation Accelerated

In the perfect condition of their flocks, in their resistance to all the complaints to which poultry are prone, in their prolific laying, they have satisfying evidence that the production by the Karswood Company of an *insectile* Spice was, for the poultry-keeper, an achievement of great moment.

"The Little More"

To goose-breeders—new and old —we say: To give Karswood Poultry Spice to geese is to give them " that little more " which can mean all the difference between ordinarily good geese, and geese that are *perfectly grand*.

All Corn Dealers sell Karswood Poultry Spice in packets 7½d. and 1/3 and larger bulk sizes.

87

KILL THOSE

INSECTS

Simple, easy and safe to use—just a pinch or two on each bird and a light dusting of nest boxes and perches is all that is necessary.

Lice and red mite are the biggest enemy of stock of all ages. They stunt the growth of chicks; worry, torment and drain away the life-blood of pullets and seriously impair the production of laying stock.

Be rid of them without a further moment's delay — Sodium Fluoride means instant death to poultry pests.

WITH

SODIUM FLUORIDE

PRICE: 10/- per 7 lbs. including postage

CASH WITH ORDER TO :—

THE NUTFIELD MANUFACTURING CO. LTD.
King's Mill Works, South Nutfield, Surrey

Books for Poultry Keepers

DOMESTIC POULTRY KEEPING (Fifth Edition). Covers all the elementary principles of poultry keeping for domestic purposes. Fully illustrated. Price 1/-, by post 1/2.

NATURAL HATCHING AND REARING (Second Edition). By C. G. MAY. Fully illustrated. Price 1/-, by post 1/2.

BANTAMS FOR EGGS. by C. G. MAY. Contents : Housing and General Management—Breeds and their Varieties—Feeding—Breeding and Chick Rearing—Hints and Common Ailments. Price 1/-, by post 1/2.

CHICK MANAGEMENT—From the Day-Old Stage (Third Edition). By I. W. RHYS, N.D.P. Describes in detail all common brooding systems. Feeding schedules, etc. Price 1/-, by post 1/2.

STARTING POULTRY KEEPING (Sixth Edition). By the Editorial Staff of "Poultry World" for war-time poultry keepers, with reference to emergency conditions. Fully illustrated. Price 2/6, by post 2/8.

POULTRY BREEDING AND PRODUCTION (Third Edition). By W. POWELL-OWEN. Covers every phase of breeding, including Waterfowl, Turkeys, Bantams and Guinea Fowl. 148 Pages. 68 Illustrations. Price 2/6, by post 2/8.

LAYING CAGES AND BATTERIES—Indoor and Outdoor (Second Edition). By W. POWELL-OWEN. The only publication of its kind on this intensive method of egg production. Separate supplement on War-time Feeding. Fully illustrated. Price 2/6, by post 2/8.

PRACTICAL POULTRY KEEPING (Second Edition). By the Advisory Staff of "Poultry World." Contents include : Laying the Foundations ; Houses and appliances ; Foods and feeding ; Breeding ; Incubation ; Chick rearing ; Sexing chicks ; Sex-linkage and auto-sexing ; Ducks ; Geese and turkeys ; Ailments ; Table poultry ; By-products and sidelines, etc. Stiff boards, 73 illustrations. Price 5/-, by post 5/3.

DUCKS (Second Edition). By R. APPLEYARD. A comprehensive work on Duck keeping by an acknowledged authority. Deals with Rearing, Breeding, Foods and Feeding, Housing, Diseases and Ailments, etc. Fully illustrated. Price 5/-, by post 5/4.

GROW YOUR OWN EGGS or Silage from your Garden. By ARTHUR SMITH. Detailed instructions on conserving summer crops for winter feeding. Construction of home-made silos, suitable crops, filling instructions and information on feeding, etc. Price 6d., by post 7d.

Obtainable through any newsagent, or direct by post from
The Publishers : POULTRY WORLD, Dorset House, Stamford St., S.E.1

www.ingramcontent.com/pod-product-compliance
Lightning Source LLC
Chambersburg PA
CBHW031440270326
41930CB00007B/804